PHYSICS OF DIELECTRICS

for the Engineer

FUNDAMENTAL STUDIES IN ENGINEERING

FUNDAMENTAL STUDIES IN ENGINEERING 1

PHYSICS OF DIELECTRICS

for the Engineer

ROLAND COELHO

Maitre de Recherche au C.N.R.S.
Ecole Supérieure d'Electricité
Gif-sur-Yvette, France

ELSEVIER SCIENTIFIC PUBLISHING COMPANY
Amsterdam – Oxford – New York 1979

ELSEVIER SCIENTIFIC PUBLISHING COMPANY
335 Jan van Galenstraat
P.O. Box 211, 1000 AE Amsterdam, The Netherlands

Distributors for the United States and Canada:

ELSEVIER/NORTH-HOLLAND INC.
52, Vanderbilt Avenue
New York, N.Y. 10017

Library of Congress Cataloging in Publication Data

Coelho, Roland.
Physics of dielectrics for the engineer.

(Fundamental studies in engineering ; v. 1)
Bibliography: p.
Includes index.
1. Dielectrics. I. Title. II. Series.
QC585.C58 537.2'4'0246213 78-10850

ISBN 0-444-41755-9 (Vol. 1)
ISBN 0-444-41756-7 (Series)

Printed in The Netherlands

PREFACE

For twelve years, I have been in charge of a course on the physics of dielectrics for electrical engineering students of the Ecole Supérieure d'Electricité in Paris. The successful reception of the course, together with the lack of an adequate textbook, has encouraged me to write this manuscript, based on my lecture notes.

I have attempted to be faithful to the outstanding teaching I received from Arthur von Hippel when I was a graduate student at M.I.T. Accordingly, my purpose was to try to teach how dielectrics might behave, and to compare this ideal behaviour with their actual behaviour.

To succeed, I had to motivate my students to apply themselves to a field that is sometimes regarded as purely technological, and remote from scientific excitement. I had to help them to dig into their background of electrostatics, wave theory, thermodynamics and statistical mechanics, to assemble the old pieces together, and to build from them a practical discipline based on a proper understanding of "molecular engineering", an expression coined by von Hippel. To assist the reader in achieving this end, many problems are set for solution. These problems, tested by my students and selected for their educational value, constitute an essential part of the text.

Although the present book is written primarily for graduate students and engineers interested in materials science, it should also be useful to technicians looking for a simple approach to the electrical properties of matter. I hope that professional scientists will also find it a worth while document, since its level is as advanced as the relatively elementary formalism used throughout allows.

Most of the material covered here is treated thoroughly in other publications, but a unique character of this book is its systematic attempt to clarify and correlate advanced concepts, to make them accessible to non specialists, and thereby fill the gap between the textbook level and that of advanced monographs.

In Part 1, a concise review of the basics of electrostatics permits a discussion of various models for the polarizability of atoms and molecules, and relates this to the macroscopic permittivity.

Part 2 presents the behaviour of matter in an alternating field, in terms of the complex permittivity, and discusses the various interactions between field and matter.

Part 3, devoted to the dissipative effects under high fields, is based on the wide-gap semi-conductor model, from which the various types of charge carriers are discussed. Finally, the main disruptive processes are described, with references to experimental procedures.

The aim to write a concise book, of use to people with widely different backgrounds could not be achieved, of course, except at the sacrifice of completeness. Many important topics, such as the various kinds of electrical polarization induced by non-electrical stresses (piezo and pyro electricity), the spontaneous and permanent polarizations (ferro-electricity, electrets, etc.), the transient, thermo- and piezo-stimulated currents, and the extrinsic breakdown mechanisms such as treeing and internal discharges have deliberately been omitted, or just quoted as applications, but general references on these topics are given.

The book is intended as an introduction to these topics, and I shall feel satisfied if its reading aids their understanding or stimulates further research.

The literature cited in the bibliography refers exclusively to the major books and review articles. Systematic reference to papers published in scientific periodicals is not required here, since the reader can extract this information from the general bibliography.

Finally, I would like to express my gratitude to those who have given of their time in discussing various aspects of the manuscript. Among others, Professor A. Jonscher of Chelsea College (London) has kindly contributed to the critical analysis of the Debye relaxation model in Part 2, and Professor N. Klein of Technion (Haifa) to the discussion of the breakdown models in Part 3.

I am deeply indebted to Mrs Hervo for her skillful typing in English, to Mrs Palierne for drawing the illustrations with artistic taste, and to my coworkers who have kindly assisted me in the correction of the final proofs. In spite of their invaluable cooperation, errors have undoubtedly escaped our scrutiny, and I should be very grateful to anyone who would be so kind as to point them out to me.

Gif-sur-Yvette
July 1978

R. COELHO

LIST OF SYMBOLS

Roman characters

a	distance
A	Ampere
b	distance
B	Magnetic induction, birefringence coefficient
c	velocity of light $c = 3 \times 10^8\ ms^{-1}$
C	capacitance
d	distance
D	electric induction, diffusion coefficient
e	electron charge, ecentricity, electric vector of light
E	electric field
f	friction coefficient, frequency, reaction field factor
F	force
g	internal field factor, orientational correlation parameter
G	conductance
h	Planck's constant, Cole-Cole parameter
H	magnetic field
i	$\sqrt{-1}$, reduced current density
I	current, current density
j,J	current density
k	Boltzmann constant, absorption index
K	relative permittivity $\varepsilon/\varepsilon_o$, thermal conductivity
l,L	length
m	mass, effective induced dipole moment
M	mass
n	refractive index ($n^* = n-ik$), number density
N	concentration
p	probability
P	polarization
q,Q	electric charge
r,R	distance
s	second
S	entropy, order parameter
t	time
T	temperature
u	unit vector
U	energy

v unit vector, velocity

V potential

w,W energy

x $\cos\theta$ (Langevin's notation)

X reduced distance

y $\frac{\mu E}{kT}$ (Langevin's notation)

z $\frac{\Delta\alpha E^2}{2kT}$

<u>Greek characters</u>

α polarizability, absorption coefficient, townsend multiplication factor

β $\Delta\alpha kT/\mu^2$, spring constant, imaginary part of complex absorption coefficient

γ depolarization factor, complex absorption coefficient ($\gamma^* = \alpha + i\beta$)

Γ heat transfer coefficient, Gamma fonction

δ loss angle, reduced distance

ε absolute permittivity ($\varepsilon^* = \varepsilon' - i\varepsilon''$)

φ phase angle, potential

Φ reduced potential, decay function, work function

Ψ wave function, angle between permanent dipole and molecular axis

λ length (wavelength, Debye length), reduced length (a/r)

μ dipole moment, magnetic susceptibility, mobility

ν concentration

ω angular frequency

Ω solid angle

ϖ probability

ρ charge density

σ electrical conductivity

τ relaxation time

CONTENTS

PART 1. MATTER IN A CONSTANT ELECTRIC FIELD

I. INTRODUCTION - CONDENSED REVIEW OF ELECTROSTATICS

The purpose of this book is to offer to readers having some familiarity with classical physics a sound basis for understanding the physical concepts involved in the behaviour of insulating materials submitted to high electric fields.

By definition, an insulator is a material through which no steady conduction current can flow when it is submitted to an electric field. Consequently, an insulator can accumulate electric charges, hence electrostatic energy, and for this reason it is a dielectric, but electrical energy can also be stored by other mechanisms. For instance, the storage and release of electrostatic and electrochemical energies are compared in Table 1.

TABLE 1

	Electrostatics	Electrochemistry
Maximum stored energy density kW.h.m^{-3}	$\leqq 0.1$	$>10^2$
Maximum current (Amp)	$\sim 10^6$ (limited only by external circuit)	$\sim 10^3$
Maximum voltage (V)	10^5 n	1.5 n

n = number of elements in series

The word "dielectric", especially if it is used as an adjective, covers a wide range of materials. Any substance to which a dielectric permittivity can be ascribed, for instance by optical methods, as we shall see later, can be regarded as dielectric, at least at high frequencies. This covers all materials, including electrolytes and even metals.

That is why the realm of the present book is not limited to insulators, and why it is hoped that all those concerned with materials science (students, engineers, physicists, chemists,etc.) will find it

TABLE 2

Quantity	Symbol	Definition	No	Units (S.I.)
Electric field	$\vec{E}$	$\vec{E} = -\vec{\nabla} V$	(1)	Vm^{-1}
Conduction current density	$\vec{j}_c$	$\vec{j}_c = \sigma \vec{E}$	(2)	$QT^{-1}m^{-2} = Am^{-2}$
Dipole moment	$\vec{\mu}$	$\vec{\mu} = q\ \vec{r}$	(3)	Qm
Polarization	$\vec{P}$	$\vec{P} = \frac{\partial \mu}{\partial v}$ (dipole moment per unit volume)	(4)	Qm^{-2}
Electric induction	$\vec{D}$	$\vec{D} = \varepsilon_o\ \vec{E} + \vec{P}$	(5)	Qm^{-2}
Induction current density	$\vec{j}_D$	$\vec{j}_D = \frac{\partial \vec{D}}{\partial t}$	(6)	Am^{-2}
Total current density	$\vec{I}$	$\vec{I} = \vec{j}_D + \vec{j}_c$	(7)	Am^{-2}
Dielectric permittivity	ε	$\vec{D} = \varepsilon \vec{E}$	(8)	
		$\vec{P} = (\varepsilon - \varepsilon_o)\vec{E}$	(8')	$QV^{-1}m^{-1}$
Electrostatic energy density	w	$dw = \vec{D}.d\vec{E}$	(9)	QVm^{-3}
		isotropic material		
		$w = \frac{1}{2} \varepsilon E^2$	(9')	
Gauss's Law (corollary)		$\int_S \vec{D}.d\vec{s} = \int_V \rho\, dv$	(10)	Q
Poisson's law		$\vec{\nabla}.\vec{D} = \rho$	(11)	Qm^{-3}
Laplace-Maxwell eqns.(without H terms)		$\vec{\nabla}.\vec{I} = 0$ using (7) and (11)	(12)	Am^{-3}
Charge conservation eqn.		$\frac{\partial \rho}{\partial t} + \vec{\nabla}.\vec{j}_c = 0$	(13)	Am^{-3}

worth reading.

The background of electrostatics required for a good understanding of the text is summarized in Table 2.

Since the advent of circuit microminiaturization, capacitors have tended to become the most expensive and bulky electronic components. This situation has generated a revival of research on dielectrics, directly or indirectly oriented towards an improvement of the ability of materials to store electrostatic energy and release it at will.

Therefore, the basic relations of Table 2, around which the present book is elaborated, are equation (9) on electrostatic energy density, and its integrated form (9'):

$$w = \frac{1}{2}\,\varepsilon E^2$$

The last equation contains only two quantities, ε and E.

The first factor ε is the absolute permittivity of the material, related to the absolute permittivity ε_o of vacuum, by $\varepsilon = K\varepsilon_o$, where the dimensionless factor K is known as the relative permittivity of the material.

E is the magnitude of the local electric field at the point under consideration. As we shall see, this field may be quite different from the applied field, for various reasons.

Consequently, the two factors of equation (9'), namely the permittivity ε, related to the microscopic properties of the materials, and the highest field the materials can withstand (or dielectric strength) will be discussed separately.

Since the first quantity often involves steady d.c. properties but is usually measured with an alternating field, it is necessary to discuss first its steady value, the static permittivity, and then its behaviour with an a.c. applied field, namely the complex permittivity.

This yields naturally the logical structure of the book:

PART 1 Matter in a constant electric field
PART 2 Matter in an alternating field
PART 3 Dissipative effects under high fields

II. THE POTENTIAL OF A GROUP OF CHARGES

Historically, the relation $F = Q^2/r^2$ giving the magnitude of the repulsion (Dyne) between two identical point charges Q separated by a distance r (cm) was purely empirical, and defined the c.g.s. electrostatic unit of charge.

This famous inverse square law was put on a firm ground by mathematical arguments (Gauss's theorem). The potential created by a point charge Q at a distance r varies as 1/r.

In the International System now in use, the unit of charge is the Coulomb, which is the charge carried by 1 Ampere during 1 second. The relation between the Coulomb, which is electromagnetic in nature, and the electrostatic c.g.s. unit involves the velocity of light (cf. VI.1). With $c = 3 \times 10^8\ ms^{-1}$, one Coulomb equals 3×10^9 u.e.s.c.g.s.

The potential (Volt) at a distance r from a charge Q (Coulomb) takes the well known form :

$$V = \frac{Q}{4\pi\varepsilon_o r} \qquad \text{(II.1)}$$

where ε_o, the permittivity of a vacuum, is $10^{-9}/36\pi$ $(F.m^{-1})$.

All constituents of matter (atoms, molecules, crystals, etc.) are collections of charges of various types : point charges according to the classical models of nuclei and electrons, or distributed charges according to quantum mechanics.

All the manifestations of the electrical properties of matter, namely all the signals by which matter reveals its existence, imply the potential created by its constituent charges, and therefore it is of fundamental importance to know how to deal with these potentials.

The two propositions given below without sophisticated demonstrations will be clarified by working out simple examples. This will, it is hoped, ease the understanding of rather abstract concepts.

II.1. Multipolar expansions

Proposition 1

Any group of charges, for instance the charges in a molecule, can be regarded from a point M far enough from the group, as a superposition of point multipoles, namely:

- a point monopole Q_o (which is a scalar)
- a point dipole $\overline{Q_o}$ (which is a vector)
- a point quadrupole $\overline{\overline{Q_2}}$ (which is a second rank tensor)
....

.....

- a point (2^n) pole Q_{2n} (which is a 2^n rank tensor).

Although we have identified in the list the tensorial nature of the multipoles, it is not necessary to be familiar with tensors to understand what follows.

We can simply consider the potential $V(M)$ at point M due to the charges as an expansion:

$$V(M) = \sum_{i=o}^{i=\infty} V_{2^i}$$

$$= V_o + V_1 + V_2 + \ldots + V_{2^i} + \ldots$$

where

V_o is the potential of the point monopole, varying with the distance as r^{-1}

V_1 is the potential of the point dipole, varying with the distance as r^{-2}

V_2 is the potential of the point quadrupole varying with the distance as r^{-3}

etc.

Proposition 2

The multipolar expansion of $V(M)$ depends on the choice of the origin.

Propositions 1 and 2 are illustrated in the following example.

Example

Let us consider a pair of charges of values (+4) and (+1) at points A and B respectively, separated by a distance 2d, and a point M far from A and B.

In order to calculate the potential $V(M)$ at M, we have to choose an origin. This origin may be taken anywhere, but it is obvious that one of the simplest choices is the mid-point O between A and B.

Another simple choice is the barycentre Ω of the charges.

(a) If we adopt O as the origin, we have:

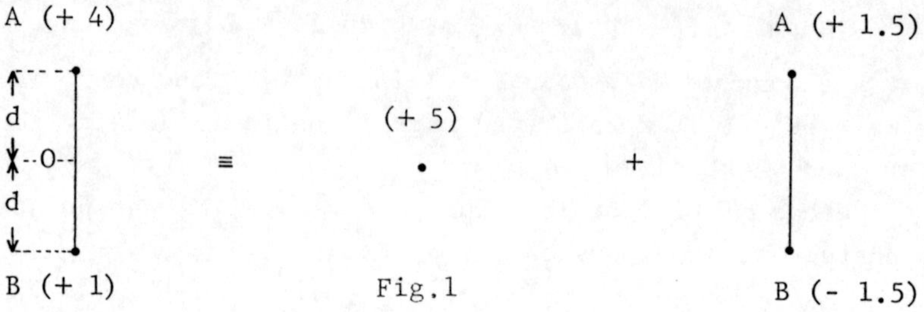

Fig.1

The charge (+5) at O is the <u>total</u> charge of the system. It is a <u>monopole</u>, and its value obviously does not depend on the origin.

The couple of equal and opposite charges ($\pm$ 1.5) constitutes a <u>real</u> dipole, and it will be shown later that a real dipole centred at the origin can be regarded as a point dipole of the same dipole moment 3d, to which must be added higher order multipoles starting with an <u>octupole</u>. Therefore:

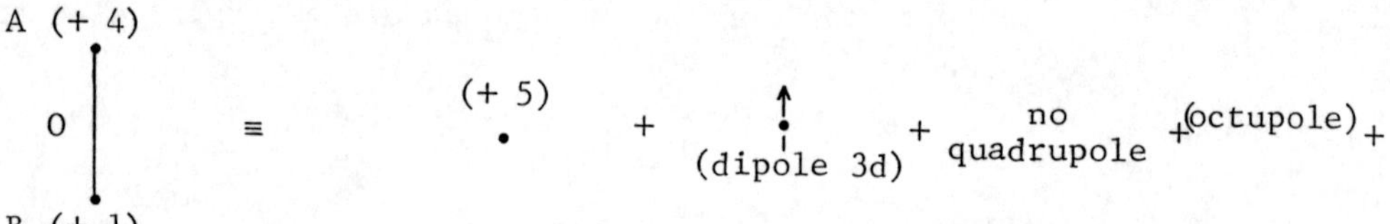

Fig.2

(b) If we adopt Ω as the origin, we have:

$$(+4)\,\overrightarrow{\Omega A} + (+1)\,\overrightarrow{\Omega B} = 0$$

which shows that the dipole moment, exprimed by $\sum q\vec{r}$, is zero at Ω . Hence, the multipolar expansion, with respect to Ω , of the two charges does not contain any dipolar term. This is confirmed by Fig.3:

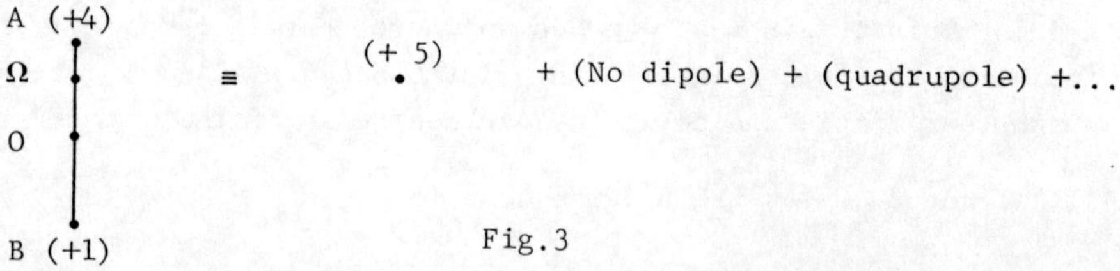

Fig.3

The vanishing dipole moment here is related to the choice of the origin at the centre of charges, which lies at a finite distance only if the monopole exists ($\sum q \neq 0$).

The statements will now be developed on a more quantitative ground, starting with the mathematical derivation of the multipolar expansion of the potential due to a single point charge.

II.2. Multipolar expansion of a single point charge

Let us consider a group of point charges and, among these charges, the point charge Q_i at A. (Fig.4)

The potential at a point M such that $|\overrightarrow{AM}|$ $|\overrightarrow{OA}| = a$ is

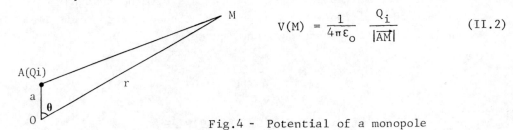

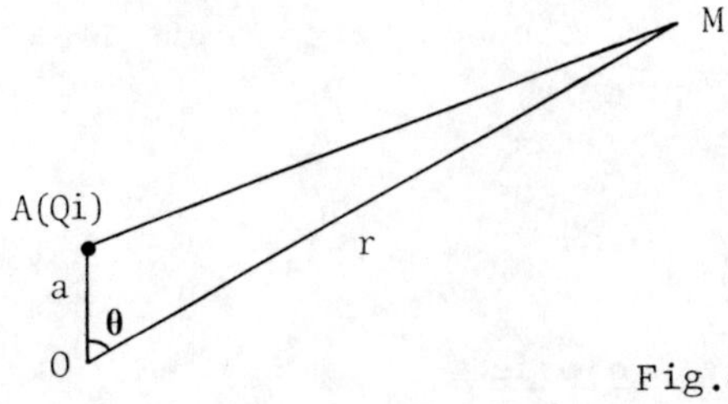

$$V(M) = \frac{1}{4\pi\varepsilon_o} \frac{Q_i}{|\overrightarrow{AM}|} \tag{II.2}$$

Fig.4 - Potential of a monopole

In the triangle OAM,

$$\overline{AM}^2 = \overline{OM}^2 + \overline{OA}^2 - 2\,\overrightarrow{OM}.\overrightarrow{OA} \tag{II.3}$$

Now, let $|\overrightarrow{OM}| = r$ $(\overrightarrow{OM}, \overrightarrow{OA}) = \theta$ and $\frac{a}{r} = \lambda \ll 1$

Then, with respect to the origin, we can write:

$$V(M) = \frac{Q_i}{4\pi\varepsilon_o r} (1 + \lambda^2 - 2\lambda\cos\theta)^{1/2} \tag{II.4}$$

Since λ, and thus also $(\lambda^2 - 2\lambda\cos\theta)$, is small with respect to unity, the expression (II.4) can be expanded as

$$V(M) = \frac{Q_i}{4\pi\varepsilon_o r} \left[1 - \frac{1}{2}(\lambda^2 - 2\lambda\cos\theta) + \frac{3}{8}(\lambda^2 - 2\lambda\cos\theta)^2 + \ldots\right]$$

and, reordering the above equation for increasing powers of λ,

$$V(M) = \frac{Q_i}{4\pi\varepsilon_o r} \left[1 + \lambda\cos\theta + \lambda^2\left(\frac{3}{2}\cos^2\theta - \frac{1}{2}\right) + \ldots\right] \tag{II.5}$$

In other words, $V(M) = V_o + V_1 + V_2 + \ldots$, where

$V_o = \dfrac{Q_i}{4\pi\varepsilon_o r}$ is the potential of the monopole Q_i at O

$V_1 = \dfrac{Q_i a}{4\pi\varepsilon r^2} \cos\theta$ is the potential of the point dipole $\vec{\mu_i} = Q_i \, \overrightarrow{OA}$

$V_2 = \dfrac{Q_i a^2}{4\pi\varepsilon_o r^3} \left(\frac{3}{2}\cos^2\theta - \frac{1}{2}\right)$ is the potential of the point quadrupole at O

etc...

The factors involving $\cos\theta$ are Legendre Polynomials $P_n(\cos\theta)$ of the order of the multipole. These polynomials have been actually identified through the multipolar expansions, and the square root of (II.4) is their "generating function". Consequently:

$$V(M) = \frac{Q_i}{4\pi\varepsilon_o r} \sum_{n=o}^{\infty} \lambda^n P_n(\cos\theta) .$$

II.3. Multipolar expansion of a real dipole

In Fig.5, we see that

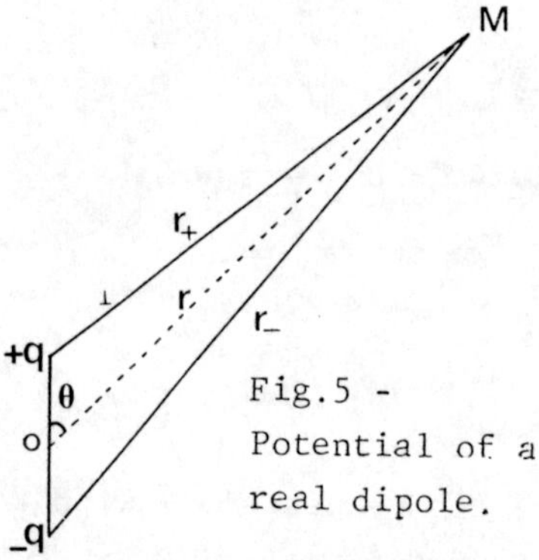

Fig.5 - Potential of a real dipole.

$$V(M) = \frac{q}{4\pi\varepsilon_o}\left(\frac{1}{r_+} - \frac{1}{r_-}\right) \tag{II.6}$$

with $r_{\pm} = r(1 + \lambda^2 \mp 2\lambda\cos\theta)^{1/2}$

r_- is deduced from r_+ by changing $\cos\theta$ into $(-\cos\theta)$. Hence, from eqn. (11.6),

$$V(M) = \frac{q}{4\pi\varepsilon_o r} \sum_{n=o}^{\infty} \lambda^n \left[p_n(\cos\theta) - p_n(-\cos\theta)\right] \tag{II.7}$$

Now, the Legendre polynomials of even order are pair: $P_{2n}(-\cos\theta) = P_{2n}(\cos\theta)$. For example:

$$P_o = 1$$
$$P_2 = \tfrac{1}{2}(3\cos^2\theta - 1)$$

and the Legendre polynomials of odd order are impair: $P_{2n+1}(-\cos\theta) = -P_{2n+1}(\cos\theta)$. For example :

$$P_1 = \cos\theta$$
$$P_3 = \tfrac{1}{2}(5\cos^3\theta - 3\cos\theta)$$

Finally, the potential of the dipole is

$$V(M) = \frac{q}{2\pi\varepsilon_o r} \sum_{n=o}^{\infty} \lambda^{2n+1} P_{2n+1} (\cos\theta) \qquad (II.8)$$

These equations confirm that the first order moment is a point dipole, hence a three-component vector, that the second order moment does not exist, that the third order moment is a point octopole, hence a third rank tensor, etc...

Of course, if a dipole is considered with respect to its direction, it is defined by only one quantity, which is its magnitude. In a similar way, if a quadrupole is viewed with respect to the principal axes of the molecule, it reduces to a three-component diagonal matrix, etc...

<u>Problem</u>

Consider the linear molecule CO_2 shown in Fig.6.

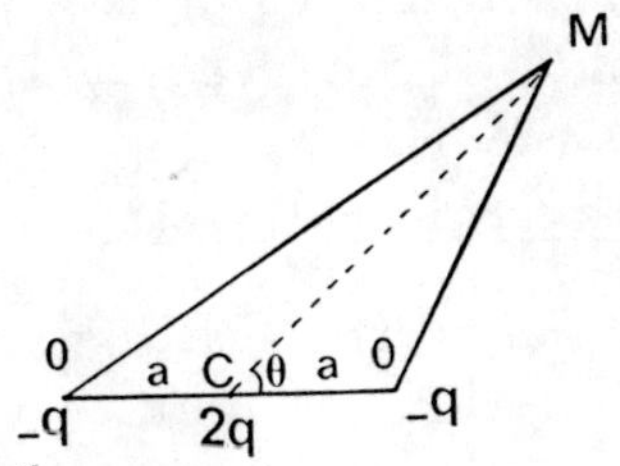

Fig.6 - CO_2 molecule

If (-q) is the charge on the oxygen sites, hence 2 q on the carbon site, show that the first term of the multipolar expansion of the CO_2 molecule at point M defined by r and θ is:

$$V(M) = \frac{-q\,a^2}{4\pi\varepsilon_o r^3} (3\cos^2\theta - 1) .$$

III. DIPOLES INDUCED IN AN APPLIED FIELD

Any particle (atom, ion, molecule) in an electric field $\vec{E}$ becomes an induced dipole moment $\vec{\mu}$. If the applied field is not too high, i.e. if it remains lower than the inner atomic or molecular field, the magnitude of the induced dipole moment $\vec{\mu}$ is proportional to that of the field:

$$\vec{\mu} = \alpha \vec{E} \tag{III.1}$$

The factor α is, by definition, the polarizability of the particle.

If the particle is spherically symmetrical, the induced moment must be in the direction of the applied field, and α is a scalar. However, particles usually have no spherical symmetry, and the induced moment in that case is not parallel to $\vec{E}$. The polarizability is a second rank tensor with respect to the principal axes of the molecule. This tensor has three characteristic values α_1, α_2 and α_3.

III.1. Quantum mechanical approach of electronic polarizability

We shall first outline the calculation of the scalar polarizability from quantum mechanics. As we shall see, this calculation is extremely complex, even in the case of simple atoms, and it is totally untractable for large molecules.

Consequently, after the short section on polarizability, we shall always use elementary classical models, which in fact yield remarkably good results, as can be expected from the correspondence principle.

It is the author's conviction, confirmed over many years of teaching experience, that it is much safer - at least for those who are not trained physicists - to deal intelligently with oversimplified models than to try to use sophisticated methods which require extensive experience before becoming productive.

This being said, consider an atomic system made of a positive nucleus surrounded by Z electrons. If the system is totally isolated, i.e. submitted to no perturbation whatsoever, the energy levels of this system are:

$$\overset{\circ}{U}_1,\ \overset{\circ}{U}_2,\ \overset{\circ}{U}_3 \ldots \overset{\circ}{U}_j,\ \overset{\circ}{U}_k \ldots$$

Now, in the presence of an electric field, which acts as a perturbation, the quantum levels above become

$$U_1,\ U_2,\ U_3 \ldots U_j,\ U_k \ldots$$

The levels perturbed by an electric field E can be obtained by use of second-order perturbation theory:

$$U_j = \overset{\circ}{U}_j - E^2 \sum_k \frac{M_{jk}}{\overset{\circ}{U}_k - \overset{\circ}{U}_j} \tag{III.2}$$

where M_{jk} is the matrix element of all the electrons, between the unperturbed states j and k , of wave functions $\overset{\circ}{\Psi}_j$ and $\overset{\circ}{\Psi}_k$:

$$M_{jk} = \sum_{i=1}^{i=z} \int \overset{\circ}{\Psi}_j \, (-ex_i) \overset{\circ}{\Psi}_k \, dv \tag{III.3}$$

In particular, the ground state $\overset{\circ}{U}_1$ becomes

$$U_1(E) = \overset{\circ}{U}_1 - E^2 \sum_k \frac{M_{1k}}{\overset{\circ}{U}_k - \overset{\circ}{U}_1} \tag{III.4}$$

From the classical standpoint, $U_1(E)$ is nothing but U_o corrected by the potential energy of the system in the field:

$$U_1(E) = \overset{\circ}{U}_1 - \frac{1}{2} \alpha E^2 \tag{III.5}$$

Simultaneous solution of (III.4) and (III.5) gives

$$\alpha = 2 \sum_k \frac{M_{1k}}{\overset{\circ}{U}_k - \overset{\circ}{U}_1} \tag{III.6}$$

The summation implies the calculation of a large number of matrix elements, each containing products of wave functions which are known analytically in only a very limited number of cases.

III.2. Elementary models for spherical atoms and molecules

Consider now three simple models.

Model 1 : The spherical negative shell with a point nucleus (Fig.7)

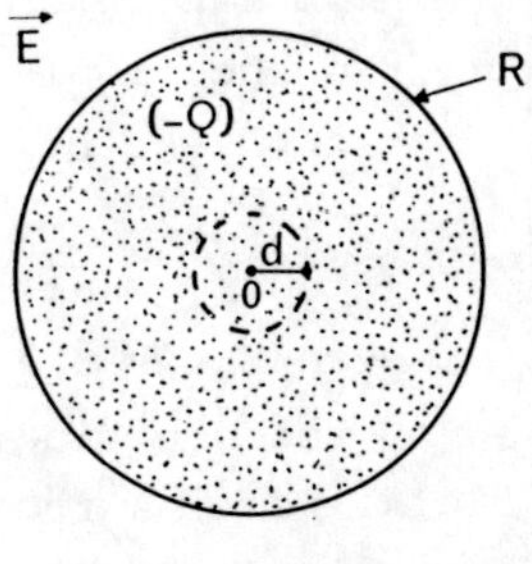

Fig.7 - The spherical atom

A neutral atom is regarded as a positive nucleus of charge (+ Q) surrounded by a spherical cloud of uniform charge density and radius R, which does not affect the permittivity ε_o of free space.

In the presence of an applied field, the central nucleus moves, in the direction of the field, to a new position at a distance d from the centre.

The induced dipole moment $\mu = Qd$ can be expressed as a function of ε_o, R and E, by equating the driving force $\vec{F} = Q\vec{E}$ and the restoring force which, from Gauss's Theorem, is due to the part $q = (d/R)^3\ Q$ of the total charge (- Q) contained in the small sphere of radius d. This gives

$$QE = \frac{Qq}{4\pi\varepsilon_o d^2} = \frac{Q^2 d}{4\pi\varepsilon_o R^3}$$

From this:

$$\mu = Qd = 4\pi\varepsilon_o R^3\ E \qquad \text{(III.7)}$$

and

$$\alpha = 4\pi\varepsilon_o R^3 \qquad \text{(III.8)}$$

It is interesting to note that the displacement d given by $4\pi\varepsilon_o R^3 E/Q$ in the case where $R = 0.5$ Å (radius of the hydrogen atom), $E = 10^8$ V m^{-1} and $Q = |e| = 1.6\ 10^{-19}$C is only of the order of 10^{-14}m, i.e. about 2×10^{-4} times the atomic radius. This result will be recalled in the chapter on breakdown.

Model 2 : Circular orbital model

Here, the atom is treated by the Bohr atomic model. A point charge (- Q) orbits on a circular trajectory around a nucleus of charge (+ Q). A field $\vec{E}$ perpendicular to the plane of the orbit displaces the nucleus from the centre of this orbit to a point M on the axis. The induced moment of the atom is then $\mu = Qd$ with $d = |\vec{OM}|$.

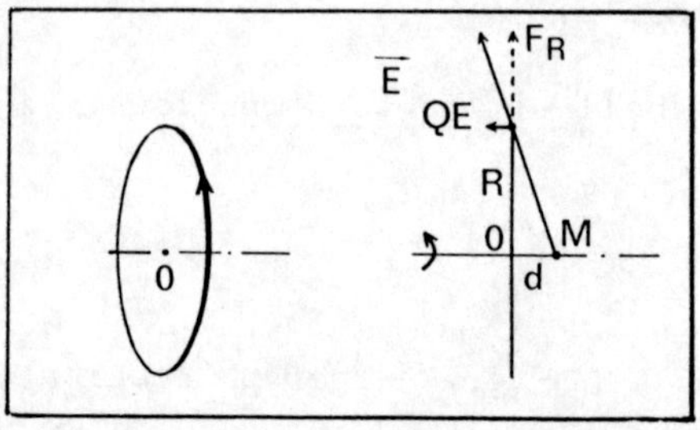

Fig.8 - The orbital model

The position of M is determined from the simple diagram in the right part of the figure :

$$\frac{d}{R} = \frac{QE}{F_R}$$

where F_R is the centrifugal force on the orbiting charge. The stable orbit, before application of the field, results from a balance between the electron-nucleus attractive force $Q^2/4\pi\varepsilon_o R^2$ and the centrifugal force (F_R).

Hence $$\frac{d}{R} = \frac{4\pi\varepsilon_o R^2 E}{Q} \tag{III.9}$$

so that

$$\mu = Qd = 4\pi\varepsilon_o R^3 E \tag{III.10}$$

This gives the same polarizability $\alpha = 4\pi\varepsilon_o R^3$ as was obtained previously (model 1).

However, if we consider a collection of similar atoms having their orbits randomly oriented rather than having all their orbits perpendicular to the field, the average induced dipole moment in the direction of the field is

$$<\mu> = \mu <\cos^2\theta> \tag{III.11}$$

If the field is not large enough to affect the average orientation, $<\cos^2\theta> = 1/3$, and

$$<\mu> = \frac{4}{3}\pi\varepsilon_o R^3 E = \varepsilon_o vE \tag{III.12}$$

where v is the volume of the sphere, so that

$$\alpha = \frac{4}{3}\pi\varepsilon_o R^3 = \varepsilon_o v \tag{III.13}$$

On the other hand, if the field is so large that all the molecular axes are parallel to the field, $<\cos^2\theta> = 1$, and $\alpha = 4\pi\varepsilon_o R^3$.

Model 3 : Material sphere of permittivity ε Fig.9

We now treat the atom as a sphere of a material of permittivity ε . It is classical problem of electrostatics to find the expression for the potential outside such a sphere, using Laplace's equation $\nabla^2 V = 0$ with the boundary conditions that the potential V and the induction $\varepsilon\partial V/\partial r$ have no discontinuities at the interface. The result is:

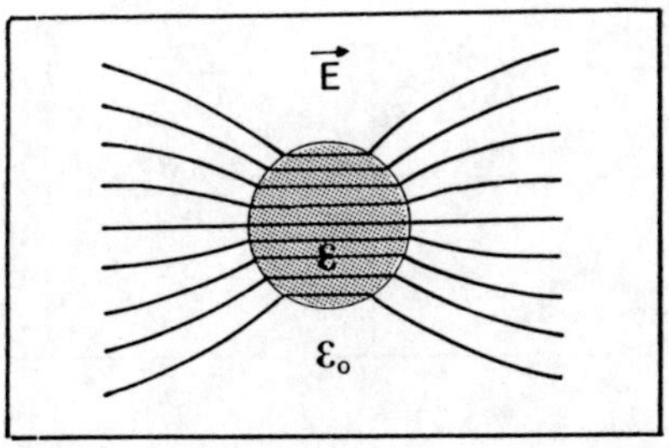

Fig.9 - The spherical ball model

$$V_{out} = \frac{\varepsilon - \varepsilon_o}{\varepsilon + 2\varepsilon_o}\frac{R^3}{r^2}\cos\theta\ E \tag{III.14}$$

showing that the sphere, viewed from its outside, behaves as a point dipole

$$\vec{\mu} = 4\pi\varepsilon_o R^3 \frac{\varepsilon - \varepsilon_o}{\varepsilon + 2\varepsilon_o}\vec{E} \tag{III.15}$$

and its polarizability α therefore is

$$\alpha = 4\pi\varepsilon_o R^3 \frac{\varepsilon - \varepsilon_o}{\varepsilon + 2\varepsilon_o} \qquad \text{(III.16)}$$

Table 3 summarizes the results of the various models, and also, in columns 4 and 5, gives the relative permittivity $\frac{\varepsilon'}{\varepsilon_o} = 1 + \frac{N\alpha}{\varepsilon_o}$ of the condensed materials made by packing the spheres of the models in two different ways : cubic and close-packed. In these results, N is the number density of the spheres. Interactions between the induced dipoles have been neglected.

TABLE 3 - The various classical models for the polarizability of spherical particles, with their respective predictions for the permittivity of condensed matter.

MODEL	Conditions	$\frac{\alpha}{\varepsilon_o v}$	Cubic packing $N=\frac{\pi}{6v}$	Close-packed $N=\frac{\pi\sqrt{2}}{6v}$
1	-	3	$1+\frac{\pi}{2} = 2.57$	$1+\frac{\pi\sqrt{2}}{2} = 3.22$
2 a	orbits $\perp \vec{E}$ (high E)	3	2.57	3.22
2 b	random orientation (low E)	1	$1+\frac{\pi}{6} = 1.52$	$1+\frac{\pi\sqrt{2}}{6} = 1.74$
3 a	$K = \varepsilon/\varepsilon_o = 2$	3/4	$1+\frac{\pi}{8} = 1.39$	$1+\frac{\pi\sqrt{2}}{8} = 1.55$
3 b	K = 3	6/5	$1+\frac{\pi}{5} = 1.62$	$1+\frac{\pi\sqrt{2}}{5} = 1.87$
3 c	K = 4	3/2	$1+\frac{\pi}{4} = 1.78$	$1+\frac{\pi\sqrt{2}}{4} = 2.10$
3 d	$K = \infty$ (metal spheres)	3	$1+\frac{\pi}{2} = 2.57$	3.22

<u>Tests of the models</u> (Table 4)

The quantum mechanical calculation of the polarizability α has been performed for a few simple ions (i.e. atoms with complete electron shells) and the results are given in column 1 of Table 4.

The simple models developed above (1, 2b, 3a, b, or c) relate α to the ionic radius R. The ionic radii derived from the calculated values of α (column 1) using models 1 and 2b are listed in columns 2 and 3, respectively.

TABLE 4 - Test of the spherical models

ion	$10^{42}\ \alpha$ from quantum mechanics eqn.(III.6)	$R=(\alpha/4\pi\varepsilon_o)^{1/3}$ (Å) (model 1)	$R=(\alpha/\pi\varepsilon_o)^{1/3}$ (Å) (model 2b)	R measured by X-ray crystallography (Å)
$O^{=}$	1.52	1.11	1.76	1.32
F^{-}	1.18	1.02	1.60	1.33
Na^{+}	0.838	0.91	1.42	1.01
Mg^{++}	0.724	0.87	1.36	0.75
$S^{=}$	2.387	1.29	2.02	1.69
Cl^{-}	1.97	1.21	1.90	1.72
K^{+}	1.48	1.10	1.72	1.30
Ca^{++}	1.287	1.05	1.65	1.02

The actual ionic radii have been measured experimentally by X-ray crystallography of ionic crystals, and the results are given in column 4 of the table.

Comparison of the measured and the calculated ionic radii demonstrates the validity of the models. The measured radii are intermediate between those deduced from model 1 and model 2b, except in the cases of divalent ions Ca^{++} and Mg^{++}, which agree well with model 1.

III.3. Elementary models for non-spherical atoms and molecules

It is feasible, although not elementary, to show that the polarizability of an elongated particle (for example an ellipsoid of revolution) depends on the direction of the applied field. The books by Smith and Durand and the encyclopedic article by Fueller-Brown give detailed calculations on this problem, showing that the polarizability along the long axis exceeds the polarizability along the short axis.

Here, we shall simply verify this result for a molecule made of two identical atoms (Fig.10). Any cigar-shaped molecule may be regarded as a superposition of diatomic molecules, so this result is of general validity.

Let us assume that each of the two atoms which constitute the molecule conforms to one of the models described above. The atom has a polarizability α proportional to its volume.

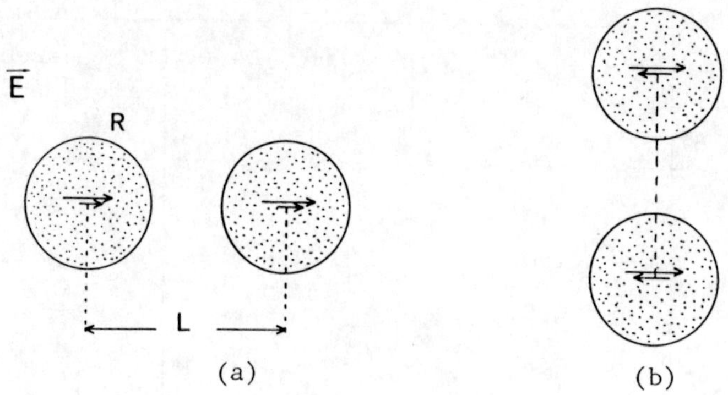

Fig.10 - The diatomic molecule
(a) molecule // to E
(b) molecule ⊥ to E

In configuration (a) of the figure, where the field is parallel to the molecular axis, each atom experiences the applied field <u>and</u> the field of magnitude $\mu / (2\pi \varepsilon_o L^3)$, due to the induced dipole $\vec{\mu}_1$ in the other atom.

If α is the polarizability of each atom, we have

$$\vec{\mu}_1 = \alpha \left(\vec{E} + \frac{\vec{\mu}_1}{2\pi \varepsilon_o L^3} \right) \qquad \text{(III.17)}$$

which can be solved for $\vec{\mu}_1$

$$\vec{\mu}_1 = \frac{\alpha}{1 - \dfrac{\alpha}{2\pi \varepsilon_o L^3}} \vec{E} \qquad \text{(III.18)}$$

Hence, the molecular polarizability in this configuration is

$$\alpha_1 = \frac{2\alpha}{1 - \dfrac{\alpha}{2\pi \varepsilon_o L^3}} > 2\alpha \qquad \text{(III.19)}$$

If we choose for α any of the models 1, 2a or 3d, for which $\alpha = 4\pi \varepsilon_o R^3$, α_1 takes the form

$$\alpha_1 = \frac{2\alpha}{1 - 2(R/L)^3} \qquad \text{(III.20)}$$

On the other hand, in configuration (b), where the applied field is perpendicular to the molecular axis, each atom again experiences the applied field, and a dipole field of magnitude $\mu_2(4\pi\varepsilon_o L^3)^{-1}$, which is now antiparallel to the applied field.

Consequently, from the same arguments as above, the molecular polarizability in configuration (b) is

$$\alpha_2 = \frac{2\alpha}{1+\dfrac{\alpha}{4\pi\varepsilon_o L^3}} < 2\alpha \qquad \text{(III.21)}$$

and, using the same model as for α_1,

$$\alpha_2 = \frac{2\alpha}{1+(R/L)^3} \qquad \text{(III.22)}$$

From (III.20) and (III.21), it is clear that

$$\alpha_2 < \alpha_1$$

The polarizability anisotropy $\Delta\alpha = \alpha_2 - \alpha_1$ is an important molecular parameter. With our present model,

$$\Delta\alpha = 2\alpha\left[\frac{1}{1-2(R/L)^3} - \frac{1}{1+(R/L)^3}\right] \qquad \text{(III.23)}$$

Since $R/L \leqslant 1/2$, $(R/L)^3 \leqslant 1/8$, so that an acceptable first-order approximation of $\Delta\alpha$ is

$$\Delta\alpha \approx 6\alpha\left(\frac{R}{L}\right)^3 \qquad \text{(III.24)}$$

It has been shown that the polarizability of a diatomic molecule (and consequently of any elongated particle of revolution) is not a scalar but a tensor with two principal values α_1 along the long axis and $\alpha_2 < \alpha_1$ along the small axis. It now becomes intuitively obvious that a particle that can be approximated as a general ellipsoid with three principal axes has, with respect to these, three polarizabilities: α_1 along the long axis, $\alpha_3 < \alpha_1$ along the short axis and α_2 $(\alpha_3<\alpha_2<\alpha_1)$ along the intermediate principal axis. This is discussed in the following problem:

o

Problem

Generalization for a general ellipsoid.

Under a field $\vec{E}$ with components E_1, E_2 and E_3, a particle acquires an induced dipole moment $\vec{\mu}$ with components μ_1, μ_2 and μ_3 where

$$\mu_1 = \sum_{i=1}^{3} \alpha_{1i} E_i \qquad \text{(III.25)}$$

with similar relations for μ_2 and μ_3.

In condensed form, (III.25) can be written as:

$$\vec{\mu} = \alpha \vec{E}$$

where α is a 3 x 3 = 9 element square matrix.

If $\vec{E}$ is changed by $\overrightarrow{dE}$, the potential energy w of the induced dipole changes by

$$dw = \vec{E}.\overrightarrow{d\mu} = \sum_{i=1}^{3} E_i d\mu_i \qquad (III.26)$$

1. Show that dw can be written

$$dw = \sum_{j=1}^{3} \sum_{i=1}^{3} \alpha_{ij} E_i \, dE_j \qquad (III.27)$$

dw must be an exact differential. Show that this condition

$$\frac{\partial}{\partial E_k}\left(\sum_i \alpha_{ij} E_i\right) = \frac{\partial}{\partial E_j}\left(\sum_i \alpha_{ik} E_i\right) \qquad (III.28)$$

reduces to $\alpha_{ji} = \alpha_{ij}$ and two similar relations. The matrix α is symmetrical and contains only six elements.

2. Show that, with a proper choice of the coordinate axes, (i.e. the principal axes of the equivalent ellipsoid), the non-diagonal matrix elements vanish, leaving a diagonal matrix

$$\alpha = \begin{pmatrix} \alpha_1 & 0 & 0 \\ 0 & \alpha_2 & 0 \\ 0 & 0 & \alpha_3 \end{pmatrix}$$

Direction and potential energy of the induced dipole in the case $\alpha_3 = \alpha_2$

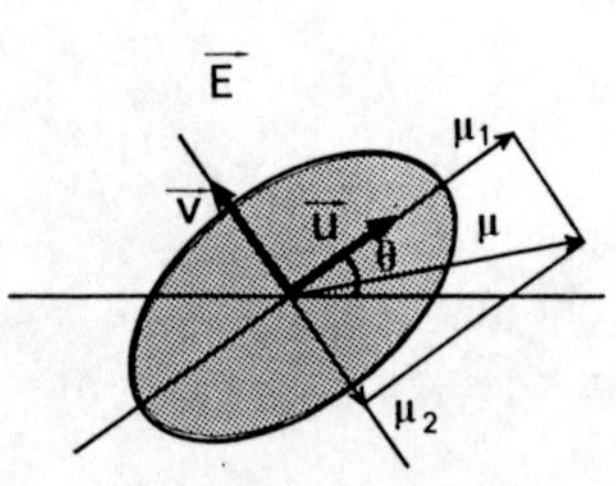

Fig.11 - Induced dipole in a long molecule

The direction and the energy of the dipole induced by an applied field on an anisotropic molecule may be found by resolving the applied field into components along the principal axes, and using the superposition principle of electrostatics (Fig.11). For a molecule of revolution, the plane of the figure is that defined by the applied field and the long molecular axis, and:

$$\vec{\mu} = \vec{\mu}_1 + \vec{\mu}_2 = \alpha_1\vec{E}_1 + \alpha_2\vec{E}_2 \tag{III.29}$$

Here,

$$\vec{E}_1 = |\vec{E}| \cos\theta\, \vec{u}$$
$$\vec{E}_2 = -|\vec{E}| \sin\theta\, \vec{v}$$

where $\vec{u}$ and $\vec{v}$ are the unit vectors on the principal axes.

The projection along $\vec{E}$ of the induced moment is:

$$|\vec{\mu}_E| = |\vec{\mu}_1| \cos\theta + |\vec{\mu}_2| \sin\theta \tag{III.30}$$

so that:

$$\vec{\mu}_E = (\alpha_1 \cos^2\theta + \alpha_2 \sin^2\theta)\vec{E} \tag{III.31}$$

This quantity will be referred to again, later.

The potential energy of the dipole induced on a spherical atom of polarizability α by an external field E is $w = -\frac{1}{2}\alpha E^2$. The factor $\frac{1}{2}$ comes from the fact that w is obtained by integrating $dw = -\mu_1 dE = -\alpha E dE$ between 0 and E.

Thus, the total energy of the particle is:

$$w = -\frac{1}{2}(\alpha_1 E_1^2 + \alpha_2 E_2^2) = -\frac{1}{2}(\alpha_1\cos^2\theta + \alpha_2\sin^2\theta)E^2 \tag{III.32}$$

Equation (III.32) can also be written in the form

$$w = -\frac{1}{2}(\Delta\alpha \cos^2\theta + \alpha_2)E^2 \tag{III.33}$$

which shows that the angular dependence of w is related to the anisotropy $\Delta\alpha$. If $\Delta\alpha = 0$, w is independent of θ, hence the particle in the field experiences no orienting torque. If $\Delta\alpha > 0$, w is a function of θ, and the particle in the field experiences a torque which tends to minimize w. From the form of $w(\theta)$, it can be seen, in the absence of interactions with other dipoles, the particular values $\theta = 0$ and $\theta = \pi$ minimize $w(\theta)$ and represent stable positions, whereas $\theta = \frac{\pi}{2}$, for which $w(\theta)$ is maximum, is an unstable equilibrium position.

Anisotropic molecule with a permanent dipole moment

For simplicity, we assume that the permanent dipole moment $\vec{\mu}_p$ is oriented along the long axis of the molecule. This is a common situation, in practice.

The total potential energy of the molecule is now

$$w(\theta) = -\vec{\mu}_p.\vec{E} - \frac{1}{2}(\Delta\alpha\cos^2\theta + \alpha_2)E^2 \quad \text{(III.34)}$$

or

$$w(\theta) = -\mu_p E\cos\theta - \frac{1}{2}(\Delta\alpha\cos^2\theta + \alpha_2)E^2 \quad \text{(III.35)}$$

where μ_p and E in (III.35) are the magnitudes of $\vec{\mu}_p$ and $\vec{E}$.

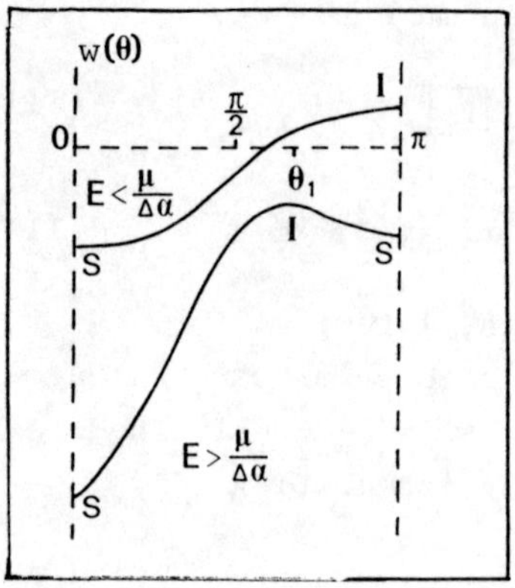

Fig.12 - w(θ) for a polar molecule

The θ dependence of w has now lost its parity in $\cos\theta$, and the variation of w with θ, shown Fig.12, depends on the sign of $(\mu_p - E\Delta\alpha)$. The derivative $dw/d\theta$, which vanishes at equilibrium, is :

$$\frac{dw}{d\theta} = E\sin\theta\,(\mu_p + E\Delta\alpha\cos\theta) \quad \text{(III.36)}$$

The $\sin\theta$ term in (III.36) implies that $\theta = 0$ and π are equilibrium positions. $\theta = 0$ is always a stable position, but for $\theta = \pi$, the stability depends on the sign of the term $(\mu_p - E\Delta\alpha)$. Thus, the value θ_1 defined by $\cos\theta = \mu_p/E\Delta\alpha$ is also an equilibrium, which exists only if $\mu_p \leqslant E\Delta\alpha$. This is summarized in Table 5.

TABLE 5 - Equilibrium of a dipolar, anisotropic molecule

θ \ E	$E < E_o$	$E_o = \frac{\mu_p}{\Delta\alpha}$	$E > E_o$
0	stable	stable	stable
θ_1	does not exist	$\theta_1 = \pi$ metastable	unstable
π	unstable		stable

Problems

The problems given below deal with the calculation of total molecular dipole moments, including the contribution of the induced dipoles, for three common molecules: ClH, H_2O and CO_2.

1. The $Cl^- H^+$ molecule - The molecule is regarded as a chlorine ion of radius R with a proton at a distance $L > R$ from the centre of the chlorine ion.

   The polarizability of the chlorine ion is assumed to conform to the spherical model ($\alpha = 4\pi \varepsilon_o R^3$).

   Show that the molecular dipole moment is $\mu = eL \times \left(1 - \frac{R^3}{L^3}\right)$

2. The water molecule - The water molecule is regarded as a spherical oxygen ion $O^=$ of center Ω and radius R with two protons at a distance $L > R$ from Ω. If the angle $(H\Omega H)$ is 2θ, show that the molecular dipole moment is:

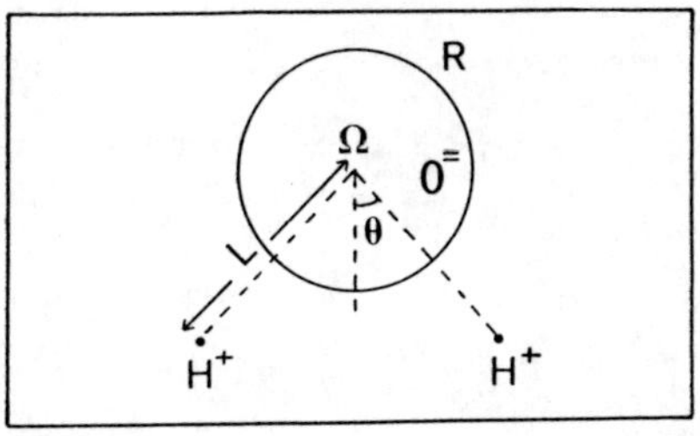

$$\mu = 2\ eL \cos\theta \left(1 - \frac{R^3}{L^3}\right)$$

Fig.13 - The water molecule

3. The CO_2 molecule - This molecule has already been studied. It was shown that the potential at a remote point $M(r,\theta)$ is:

$$V_M = \frac{q\ a^2}{4\pi \varepsilon_o r^3} (1 - 3 \cos^2\theta)$$

   Show that, if the polarizability α_o of the oxygen ions, of radius R, is taken into account, the potential V_M becomes

$$V'_M = \frac{1 - 13\ u}{1 + u} V_M$$

where
$$u = \frac{\alpha_o}{16\pi\varepsilon_o a^3} = \frac{R^3}{4a^3}$$

Of course, the models used in these problems are only very rough approximations to the real molecules; they serve only to illustrate the text.

III.4. Harmonic oscillator model for the ionic polarizability

Consider a molecular dipole qd as shown in Fig.14. In the presence of a field $\vec{E}$ parallel to the dipole, distance d increases to (d + x), where x is given by the force balance equation:

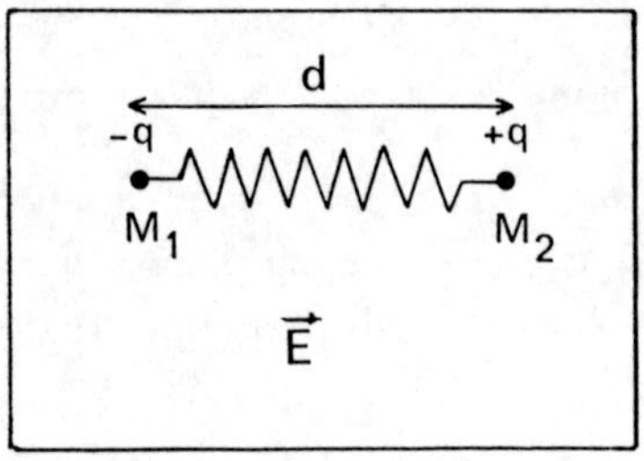

Fig.14 - The harmonic oscillator model .

$$qE = \beta x,$$

Coefficient β is the spring constant.

The dipole moment also increases by an amount

$$\Delta\mu = qx = \frac{q^2}{\beta} E,$$

and the corresponding polarizability becomes $\alpha = q^2/\beta$. With the reduced mass $m = (\frac{1}{M_1} + \frac{1}{M_2})^{-1}$, the dynamical equation of the oscillator takes the form:

$$\frac{d^2x}{dt^2} + \frac{\beta}{m} x = 0;$$

From this equation, the eigen frequency of oscillation is:

$$f = 2\pi\omega = 2\pi\sqrt{\beta/m}$$

This relates the spring constant β and the frequency ω ($\beta = m\omega^2$), so that the ionic polarizability becomes:

$$\alpha_i = \frac{q^2}{m\omega^2}$$

When typical values for the atomic masses and infrared absorption frequencies are used, the resulting values of α_i are of the order of 10^{-38} F m^2.

Problem

The unidimensional lattice.

This problem deals with the simplest model of a polar crystal, consisting of a unidimensional lattice of alternate positive and negative ions with charges (+q) and (-q) respectively.

PART I - Ionic polarizability

The ions are regarded as point charges. In the presence of an external field oriented along the chain, the positive ions move to the right and the negative ions to the left, so that the "positive lattice" is shifted by an amount $u \ll a$ with respect to the negative one (Fig.15).

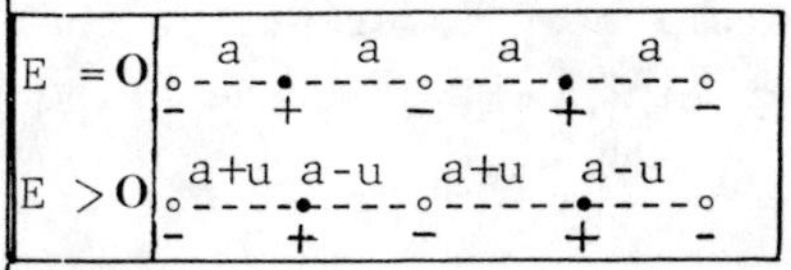

Fig.15 - The unidimensional lattice.

1. The force acting on the ions is

$$F_{\pm} = \pm \frac{q^2}{4\pi\varepsilon_0} \sum_{n=0}^{\infty} \left\{ \left[(2n+1)a-u\right]^{-2} - \left[(2n+1)a+u\right]^{-2} \right\}$$

Expand the square brackets in terms of $\eta = u/a$, and show that $F_{\pm}$ becomes

$$F_{\pm} = \pm \frac{q^2 u}{\pi\varepsilon_0 a^3} \sum_{n=0}^{\infty} \left[\frac{1}{(2n+1)^3} + \frac{2\eta^2}{(2n+1)^5} + \frac{3\eta^4}{(2n+1)^7} + \ldots \right]$$

which reduces to

$$F_{\pm} \simeq \pm \frac{q^2 u}{\pi\varepsilon_0 a^3} (1.052 + 2\eta^2 + 3\eta^4 + \ldots)$$

2. Neglecting the very small non-linear terms in u in the above expansion, find the spring constant β of the "elastic" lattice, where $\beta u = qE$.

3. Calculate the polarization of the lattice per unit length, and the resultant polarizability. Discuss the same questions for the case where the field is perpendicular to the lattice.

PART II - Electronic polarizability of the undistorted chain

The polarizability of the ions which was ignored in part I, is now considered for a rigid non-distorted chain (u=o).

4. Find the polarizability per unit length of an infinite chain of spherical ions of radius $R \ll \frac{a}{2}$, assuming that ions of both signs have the same polarizability. Study both configurations with the field parallel and perpendicular to the chain. Use model 1 (Section

III.2) for the atomic polarizability.

5. Solve the same problem, but now assuming that the positive and negative ions have different radii $(R+\rho)$ and $(R-\rho)$, respectively, and therefore different polarizabilities. The local fields at the positive and negative ion sites are now different, so that the problem involves a system of two linear equations to describe these. Solve numerically for ions in contact $(2R = a)$.

PART III - Ionic and electronic polarizabilities

Using suitable approximations, combine the effects of ionic and electronic polarizabilities.

IV. STATISTICAL THEORIES OF DIPOLE ORIENTATION IN AN APPLIED FIELD

So far, single, isolated molecules have been considered from the stand-point of electrostatics, but nothing has been said on interacting molecules. The object of this important chapter is to study molecules in interaction, at a temperature T.

IV.1. Case of free point dipoles (Langevin's theory)

We consider a collection of molecular dipoles in thermal equilibrium at a temperature T. We assume that all the molecules are identical and that they can assume any orientation. Because of thermal energy, each molecule undergoes successive collisions from the surrounding molecules.

In the absence of an applied field, the collisions tend to maintain a perfectly isotropic statistical orientation of the molecules. This means that for each dipole pointing in one direction there is, statistically, a corresponding dipole pointing in the opposite direction. This is schematically represented by configuration I in Fig.16.

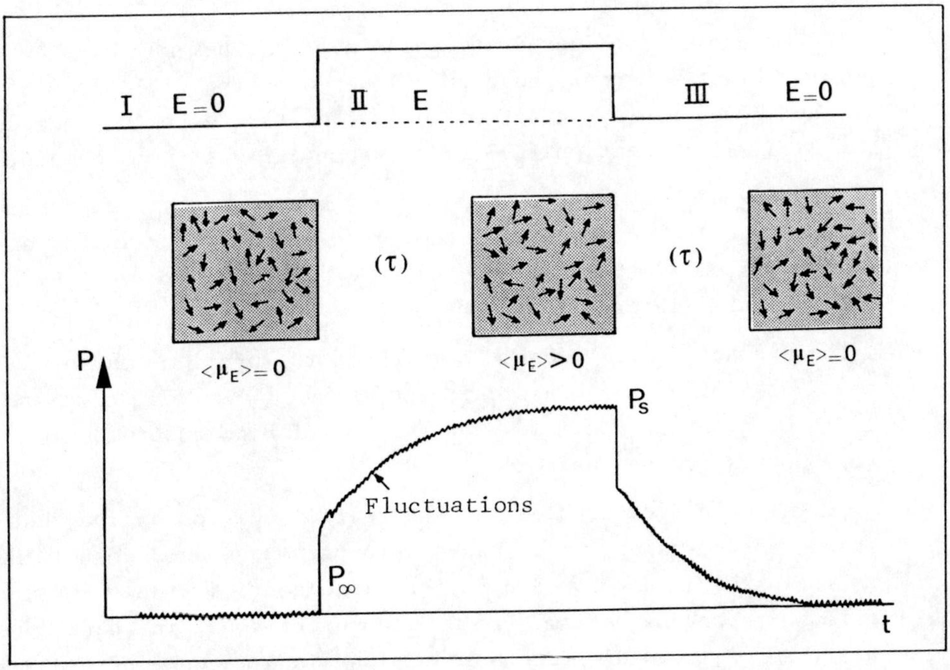

Fig.16 - Orientation of dipoles in a field

In the presence of an applied field E, the molecules experience a torque which tends to orient them parallel to the field. However, the thermal energy counteracts this tendency, and the system finally attains a new statistical equilibrium which is schematically represented by configuration II of the figure.

In this configuration, more dipoles are pointing along the field than against it. The system becomes slightly anisotropic, with a time constant τ_1 which will be the subject of the chapter dealing with dipolar relaxation, and the average component of the dipoles along the field finally assumes a steady positive value which will now be calculated.

Since the configuration has symmetry of revolution about the direction of the field, the required quantity is simply:

$$\langle\mu_E\rangle = \mu \langle\cos\theta\rangle \qquad \text{(IV.1)}$$

where θ is the angle between field and dipole, and where the angular brackets $\langle\ \rangle$, mean a statistical ensemble average, to be clarified shortly.

The calculation, due to Langevin, implies the following assumptions:

1. The molecules considered are point dipoles.They may have an isotropic polarizability; hence, the dipole induced by the field is parallel to it, and consequently this field exerts no torque on the induced dipole (cf. III.3). In section IV.3,this assumption will not be maintained.
2. The "ergodic" hypothesis holds. This means that the ensemble average taken at an instant t (i.e. the average taken over an instant "flash" of the system) is the same as the time average, taken on any element, over the time interval $(-\infty, t)$.
3. The characteristic quantum numbers of the problem are so high that the system obeys the classical statistics of Maxwell-Boltzmann, the limit of quantum statistics for systems of high quantum numbers (cf. Correspondence principle).

The molecules considered here fulfill this condition, but electron spins, for instance, which cause spin paramagnetism, do not. Hence, the Langevin theory developed here for paraelectricity does not apply in extenso to paramagnetism. Proper account of the spin quantum numbers and their related discrete orientations in a magnetic field has been taken by L. Brillouin.

According to the statistics of Maxwell-Boltzmann, the relative number of dipoles making an angle between θ and $(\theta + d\theta)$ with the applied field (i.e. pointing with the elementary solid angle $d\Omega = 2\pi \sin\theta\, d\theta$), is

$$dN(\theta) = \exp\left[-\frac{w(\theta)}{kT}\right] 2\pi \sin\theta\, d\theta \qquad \text{(IV.2)}$$

where $w(\theta)$ is the potential energy of a dipole with direction θ :

$$w(\theta) = -\mu E \cos\theta$$

Hence, the weighting function in the Boltzmann average is proportional to

$$\varphi(\theta) = \exp\left(\frac{\mu E}{kT}\cos\theta\right) \sin\theta \qquad \text{(IV.3)}$$

so that

$$<\cos\theta> = \frac{\int_0^\pi \cos\theta \exp\left(\frac{\mu E}{kT}\cos\theta\right)\sin\theta\, d\theta}{\int_0^\pi \exp\left(\frac{\mu E}{kT}\cos\theta\right)\sin\theta\, d\theta} \qquad \text{(IV.4)}$$

We now introduce the new variables :

$$x = \cos\theta\ ,$$

$$y = \frac{\mu E}{kT}$$

$$<\cos\theta> = <x> = \frac{\int_{-1}^{+1} x \exp yx\, dx}{\int_{-1}^{+1} \exp yx\, dx} \qquad \text{(IV.5)}$$

The normalizing integral of the denominator is $2\,\sinh y/y$.

The numerator can be calculated by integration by parts, or by noticing that it is the derivative with respect to y of the denominator. Finally

$$\boxed{<\cos\theta> = <x> = \coth y - \frac{1}{y} = \mathcal{L}(y)} \qquad \text{(IV.6)}$$

$\mathcal{L}(y)$ is called the Langevin function. Its graphical representation is given in Fig.17.

We see that $<\cos\theta>$ increases from 0 to 1 with increasing y (and hence with increasing E/T). This is expected since, for high values of E/T, the orienting action of the field dominates over the disorienting action of the temperature, so that all the dipoles tend to become parallel to the applied field.

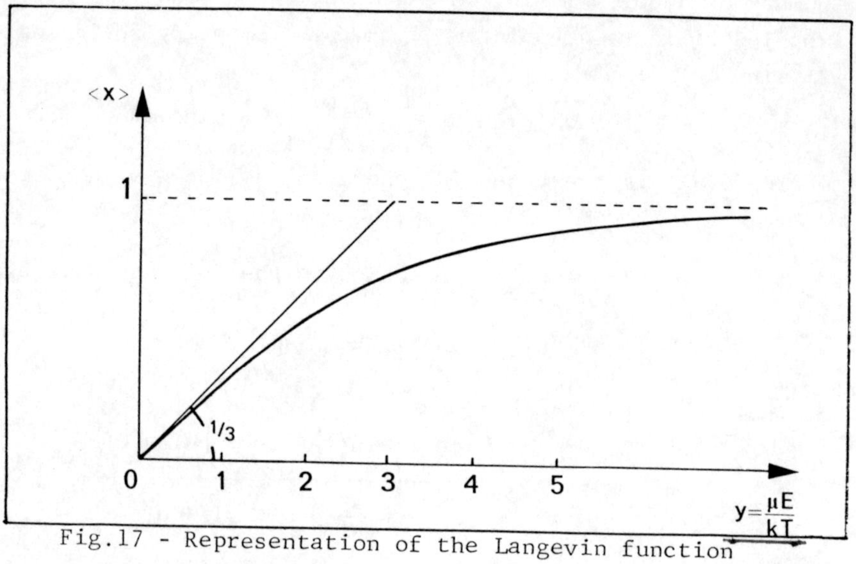

Fig.17 - Representation of the Langevin function

Practical magnitude of y

In all practical cases, y is much smaller than unity, so that only points in Fig.17 in the vicinity of the origin are of physical significance.

In order to justify this, consider (Fig.18), a dipole consisting of two charges ($\pm e$), distance 1 Å (10^{-10}m) apart, in a field $E = 10^4 \mathrm{V\,cm}^{-1}$ at ambient temperature ($kT = \frac{1}{40}$ eV). From the definition of y ,

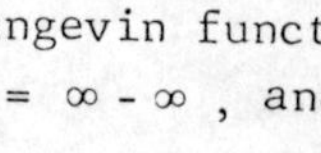

$$y = 10^{-10} \times e \times 10^6 \times \frac{40}{e} = 4 \times 10^{-3} \ll 1$$

Fig.18 - Typical dipole

Consequently, the point representing the system on the Langevin curve is very close to the origin. It is therefore necessary to expand the Langevin function about $y = 0$ to remove the indetermination $\mathcal{L}(0) = \infty - \infty$, and show that $\mathcal{L}(y) \simeq \frac{1}{3} y$.

Problem 1

Finding the expansion of $\mathcal{L}(y)$ about $y=0$ is a rather tricky problem. This difficulty can be avoided by expanding the exponentials in the integrals of (III.42) before integration, so that the integrals take the forms of simple polynomials in y. By this method, show that the third-order expansion of $\mathcal{L}(y)$ about $y=0$ is

$$\mathcal{L}(y) = \frac{y}{3} - \frac{y^3}{45} + \ldots$$

Problem 2

The Langevin theory refers to dipoles which may assume any orientation in space. This is not the case in crystal lattices, which are considered here.

(a) Dipoles in a linear crystal : In a linear lattice, dipoles μ can be only parallel or antiparallel to the field along the chain. Calculate $\langle \cos\theta \rangle$ as a function of $y = \frac{\mu E}{kT}$ about $y = 0$.

(b) Dipoles in a two-dimensional crystal : Repeat (a), for a two-dimensional lattice with a field parallel to one of the lattice directions (the dipoles may have four orientations).

(c) Dipoles in a three-dimensional crystal : Repeat (a) for a three-dimensional cubic lattice with a field parallel to one of the lattice directions. (The dipoles may have six orientations). Plot $\langle \cos\theta \rangle$ versus y on the same graph for cases (a), (b) and (c). Compare with the Langevin curve.

(d) Repeat (c), but with a field not coinciding with a lattice direction : Let α, β, and γ be the direction cosines of the field with respect to the crystal axes.

(e) Returning to (a), assume that the field acting locally on the dipoles is related to the applied field $\vec{E}$ by $\vec{E}_{loc} = \vec{E} + \gamma\vec{P}$. Show that the lattice can be spontaneously polarized in the absence of an applied field, provided that the temperature is lower than some critical temperature T_c.

Effect on the dielectric permittivity

The dipole orientation contributes to the polarization of the system, and we shall now calculate this contribution $\vec{P}_{or}$.

If N is the concentration (number per unit volume) of dipoles in the field $\vec{E}$, the orientation polarization is

$$\vec{P}_{or} = N \langle \vec{\mu}_E \rangle = N\vec{\mu} \langle \cos\theta \rangle \qquad \text{(IV.7)}$$

Since y is small with respect to unity, $\langle \cos\theta \rangle = \frac{y}{3} - \frac{y^3}{45} + \ldots$

(see problem 1) and

$$\vec{P}_{or} = \frac{N\mu^2}{3kT}\left(1 - \frac{\mu^2E^2}{15k^2T^2} + \dots\right)\vec{E}$$

Provided that the field is not too large, the bracket reduces to unity and the orientational polarization $\vec{P}_{or}$ is proportional to the field $\vec{E}$. This defines a molecular polarizability

$$\alpha_{or} = \frac{\mu^2}{3kT} \qquad \text{(IV.8)}$$

For a typical dipole $\mu = e \times 10^{-10}$ (C.m), $\alpha_{or} \approx 2 \times 10^{-38}$ (F.m^2) as compared to 10^{-40} F.m^2 for the electronic polarizability based on the models of paragraph III.2 with R = 1 Å.

Consequently, the orientational polarizability is comparable with the ionic polarizability and dominates the electronic polarizability.

Problem 3

The electrocaloric effect : a thermodynamic correction to equation (9') of Table 2.

Charging a capacitor implies an ordering of the dielectric, for instance by dipole orientation, and consequently a reduction of entropy. If the charge is isothermal, it involves an exchange of heat with the surroundings. Conversely, if the charge is adiabatic, it produces a temperature change in the dielectric.

Using E and T as independent variables, consider two neighbouring states of the charge, i.e. state 1 (E, T) and state 2 (E + dE, T + dT). The difference in internal free energy between states 1 and 2 is

$$dU = \frac{\partial U}{\partial E}\,dE + \frac{\partial U}{\partial T}\,dT$$

1. Show that the elemental variation of the electrostatic energy w between states 1 and 2 is

$$dw = \varepsilon\,E dE + E^2\,\frac{\partial \varepsilon}{\partial T}\,dT \qquad \text{(IV.9)}$$

2. Calculate the variation of entropy

$$dS = \frac{dQ}{T} = \frac{dU - dw}{T}$$

in terms of dE and dT .

3. The variation dS must be an exact differential, according to the second law of thermodynamics. Show that this condition implies:

$$\frac{\partial U}{\partial E} = E(\varepsilon + T\frac{\partial \varepsilon}{\partial T}) \qquad \text{(IV.10)}$$

4. Calculate the total change of internal free energy:

$$\Delta U = U(T,E) - U(T,0)$$

during an isothermal charge, by integrating the equation of question 3.
Show that an amount of heat ΔQ must be withdrawn from the material for it to maintain a constant temperature, and calculate ΔQ, assuming that the permittivity ε is a function only of dipole orientation:

$$\varepsilon(T) = \varepsilon_o + \frac{N\mu^2}{3kT} .$$

IV.2. Case of point dipoles in crystal lattices

Definition of a dipole lattice defect.

Consider, for example, a cation vacancy in a simple lattice such as sodium chloride. This is a typical lattice defect. Other types of more complicated structural defects (substitutional ion, interstitial, etc.) are also possible.

This cation vacancy behaves like a negative charge, and, as such, is the source of a Coulomb potential. If the lattice temperature is sufficiently high, so that the ions are relatively mobile, one of the anions in the vicinity of the cation vacancy will be expelled from the lattice by the Coulomb potential of the cation vacancy, leaving an anion vacancy which forms a dipole with the nearby cation vacancy. Here, the dipole

$\boxed{\text{Na}}^{-}\ \boxed{\text{Cl}}^{+}$

is oriented along one of the six crystallographic directions (± 1, ± 1, ± 1).

The vacancy coupling considered here is only one possible crystalline dipole, the concept of which was introduced by Breckenridge in 1954. Similar dipoles can also appear whenever an ion of different valency is substituted for the host ion. For instance, if a divalent atom such as calcium is substituted for a monovalent cation in an alkali halide, it releases two electrons and becomes a doubly charged ion (M^{++}). The new M^{++} cation has an excess positive charge which couples itself with a negative defect, such as an alkali vacancy $\boxed{\text{M}}^{-}$ or an interstitial halide, to form a dipole.

Another lattice dipole of particular historical importance occurs when Mg^{++} ions are substituted for Li^+ in lithium fluoride. To illustrate the previous sections, the dipole orientation theory of Langevin will now be applied to the magnesium-doped lithium fluoride lattice.

Dipole orientation in a crystal lattice

Fig.19 - Li^-Mg^+ dipoles in a LiF lattice

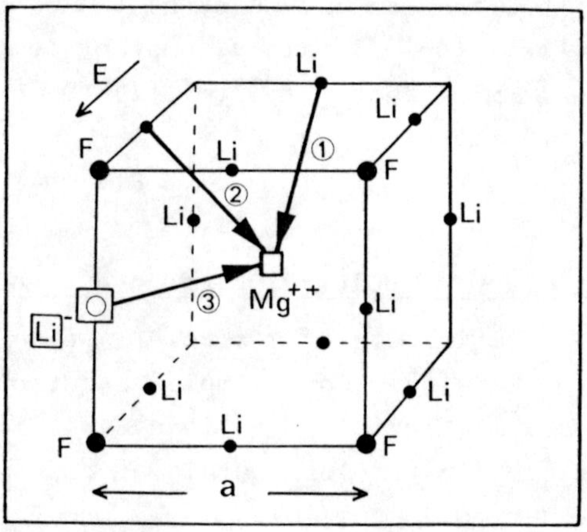

The Fig.19 represents the unit cell of crystalline lithium fluoride in which the central Li^+ ion has been replaced by a Mg^{++} ion. This substitutional ion, with its excess positive charge, forms a dipole, with a negative lithium vacancy $\boxed{Li}^-$ sitting on one of the twelve nearest-neighbour sites normally filled by a Li^+ ion. The formation of the dipole is favoured here by the small size of the Li^+ ion, which enables it to move relatively freely through the lattice away from the Mg^{++} ion.

In the absence of an electric field, these twelve sites are strictly equivalent, and the lithium vacancy hops between them giving a zero average dipole moment for the defect.

In the presence of an applied field, the twelve sites are no longer equivalent. For instance, if the field is as shown in the figure, the twelve equivalent sites of the zero-field case split into :

4 sites with lithium vacancies behind the central Mg^{++} (sites 1) and energy

$$w_1 = - \vec{\mu} \, . \, \vec{E} = - \frac{a}{2} eE \qquad \text{(IV.12)}$$

4 sites with lithium vacancies in the middle plane (sites 2) and energy

$$w_2 = 0$$

4 sites with lithium vacancies in front of the central Mg^{++} (sites 3) and energy

$$w_3 = - \vec{\mu} . \vec{E} = + \frac{a}{2} eE \qquad (IV.13)$$

The probability of a lithium vacancy occupying each type of site is different. The table below gives the energy and relative probability of occupation of each site.

Site	Energy	Boltzmann factor or (non normalized) Probability
1	$- \frac{a}{2} eE$	e^y
2	0	1
3	$+ \frac{a}{2} eE$	e^{-y}

In this table, we have used the notation $\frac{aeE}{2kT} = y$, which will be used from now throughout. The average dipole moment of the defect in the direction of the field is

$$\langle \mu_E \rangle = \frac{\sum_i \mu_i e^{-w_i/kT}}{\sum_i e^{-w_i}} \quad \text{with } i = 1,2 \text{ and } 3.$$

Hence

$$\langle \mu_E \rangle = \frac{ae}{2} \frac{e^y - e^{-y}}{e^y + 1 + e^{-y}} \qquad (IV.14)$$

Under normal field and temperature conditions, $\frac{aeE}{2kT} = y \ll 1$, implying that

$$\langle \mu_E \rangle \simeq ae \frac{y}{3} = \frac{a^2 e^2}{6kT} \times E$$

The magnitude of the dipole moment is $\mu = \frac{ae}{\sqrt{2}}$; hence

$$\langle \mu_E \rangle \simeq \frac{\mu^2 E}{3kT}, \qquad (IV.15)$$

as given by Langevin's theory.

Consequently, the contribution of the dipoles to the permittivity is again

$$\Delta \varepsilon = \varepsilon_s - \varepsilon_\infty = \frac{N\mu^2}{3kT} \qquad (IV.16)$$

IV.3. Case of polarizable dipoles with $\Delta\alpha > 0$

Hypotheses - In addition to the four assumptions of the Langevin theory, we now add the assumptions stated below:

- .The molecules are anisotropic; they have a symmetry axis, and consequently two polarizability components:
  - α_1 along the long axis
  - α_2 perpendicular to this axis
- .The direction of the permanent dipole moment μ is that of the long axis.

Notations

In addition to the notations of the Langevin theory: $x = \cos\theta$ and $y = \mu E/kT$, we use the polarizability anisotropy $\Delta\alpha$ and the average polarizability $\bar{\alpha} = (\alpha_1 + 2\alpha_2)/3$.

Furthermore, we introduce the new dimensionless quantities:

$$z = \frac{\Delta\alpha E^2}{2kT} \quad \text{and}$$

$$\beta = \frac{2z}{y^2} = \frac{\Delta\alpha kT}{\mu^2} \quad \text{(independent of E)}$$

By definition, the polarization of the material is:

$$P = N \langle\mu_E\rangle \tag{IV.17}$$

In this equation, the brackets $\langle\ \rangle$ again indicate a statistical average (here the Boltzmann average), and $\langle\mu_E\rangle$ is the projection, along the direction of applied field E , of the resultant dipole moment (permanent + induced).

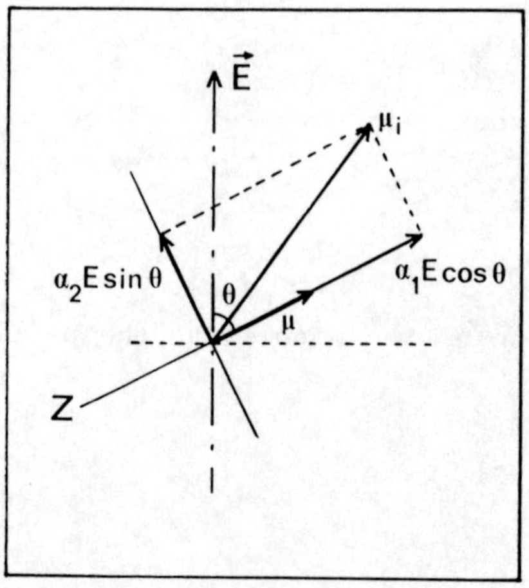

Fig.20 - Dipole moments in an anisotropic molecule. The permanent dipole $\vec{\mu}$ is assumed to lie on the long axis.

Fig.20 shows that:

$$\mu_E = \mu \cos\theta + (\alpha_1 \cos^2\theta + \alpha_2 \sin^2\theta)E$$

or

$$\mu_E = \mu \cos\theta + \Delta\alpha E \cos^2\theta + \alpha_2 E \qquad \text{(IV.18)}$$

The statistical average of the sum being the sum of the average, we have to calculate the statistical averages of $\cos\theta$ and of $\cos^2\theta$. To do this, we must know the potential energy of the polarized molecule in the applied field E.

The potential energy w_p of the <u>permanent</u> dipole is:

$$w_p = -\vec{\mu}.\vec{E} = -\mu E \cos\theta \qquad \text{(IV.19)}$$

The potential energy w_i of the <u>induced</u> dipole is:

$$w_i = -\frac{1}{2}(\alpha_1 \cos^2\theta + \alpha_2 \sin^2\theta)E^2 \qquad \text{(IV.20)}$$

The total potential energy is therefore the sum

$$w = w_p + w_i$$

The statistical average of μ_E is:

$$<\mu_E> = \frac{\int \mu_E \exp(-w/kT)\,d\Omega}{\int \exp(-w/kT)\,d\Omega} \qquad \text{(IV.21)}$$

where $d\Omega$ is the elemental solid angle around the direction of E ($d\Omega = 2\pi \sin\theta\, d\theta$). Consequently,

$$<\mu_E> = \mu <\cos\theta> + \Delta\alpha E <\cos^2\theta> + \alpha_2 E \qquad \text{(IV.22)}$$

It should be noted here that the quantity $<\cos^2\theta>$ varies from $\frac{1}{3}$ for randomly oriented molecules to 1 for the case where all the molecules are parallel or antiparallel to the field E. The quantity

$$S = \frac{3}{2} <\cos^2\theta> - \frac{1}{2} \qquad \text{(IV.23)}$$

which varies from 0 to 1 under the above conditions, represents the orientational order of the system. This quantity is called the "order parameter". Using S instead of $<\cos^2\theta>$ in the expression of $<\mu_E>$, one obtains:

$$\langle\mu_E\rangle = \mu\langle\cos\theta\rangle + \frac{3}{2}(S\,\Delta\alpha + \bar{\alpha})E \qquad \text{(IV.24)}$$

where we note that the order parameter S is associated with the polarisability anisotropy $\Delta\alpha$.

The order parameter S is used extensively in the description of mesomorphic phases such as liquid crystals.

<u>Calculation of $\langle\cos\theta\rangle$ and $\langle\cos^2\theta\rangle$</u>

1. <u>$\langle\cos\theta\rangle$</u>. By definition of the statistical average,

$$\langle\cos\theta\rangle = \frac{\int_{-1}^{+1} x\,\exp(yx + zx^2)\,dx}{\int_{-1}^{+1} \exp(yx + zx^2)\,dx} \qquad \text{(IV.25)}$$

The term in α_2 in the energy w multiplies both the numerator and the denominator of $\langle x\rangle$ and thus does not appear in the above equation. Usually, the integrals involved in the calculation of $\langle\cos\theta\rangle$ cannot be written in analytic form. However, since y and z are quite small with respect to unity in all practical cases, the exponential may be expanded to the second order in x:

$$\exp(yx + zx^2) = 1 + yx + (z + \frac{y^2}{2})\,x^2 + \ldots$$

from which the expansions of the numerator and of the denominator may be easily found. The simple algebra of the expansion of the ratio of two polynomial expansions can then be used to find the third order expansion of $\langle\cos\theta\rangle$ in y:

$$\langle\cos\theta\rangle = \langle x\rangle = \frac{y}{3}\left[1 - (1-2\beta)\,\frac{y^2}{15} + \ldots\right] \qquad \text{(IV.26)}$$

2. <u>$\langle\cos^2\theta\rangle$</u> . Using exactly the same technique as was used above for $\langle\cos\theta\rangle$,

$$\langle\cos^2\theta\rangle = \langle x^2\rangle = \frac{1}{3} + \frac{2}{45}(1+\beta)y^2 + \ldots \qquad \text{(IV.27)}$$

Note that if this expansion is limited to the second order in y, as above, it gives for the order parameter:

$$S = (1+\beta)\,\frac{y^2}{15} \qquad \text{(IV.28)}$$

3. $<\mu_E>$. Using the results obtained here for $<\cos\theta>$ and $<\cos^2\theta>$, eqn. (IV.22) becomes

$$<\mu_E> = \frac{\mu y}{3}\left[1-(1-2\beta)\frac{y^2}{15}+\dots\right]+\frac{\Delta\alpha E}{3}\left[1+\frac{2}{15}(1+\beta)y^2+\dots\right] \qquad \text{(IV.29)}$$

From the definition of $y = \mu E/kT$, the three terms involved in $<\mu_E>$ all contain E as a common factor. It therefore follows that

$$<\mu_E> = \left[\bar{\alpha}+\frac{\mu^2}{3kT}-(1-4\beta-2\beta^2)\frac{\mu^4E^2}{45k^3T^3}+\dots\right]E \qquad \text{(IV.30)}$$

In other words, deviation from the Langevin expansion appears only in the non-linear terms.

N.B. The sign of the corrected third-order term in $<\mu_E>$ depends on the value of β.

If $\beta < \beta_0 = 0.224$, the coefficient $(1 - 4\beta - 2\beta^2)$ is positive, and the polarization saturates at high fields, as predicted in the Langevin theory.

If $\beta > \beta_0 = 0.224$, the coefficient $(1 - 4\beta - 2\beta^2)$ is negative, and, in contrast to the Langevin theory, the polarization increases more than linearly with the field. This is antisaturation.

Special values of β

- $\beta = \beta_o$. The polarization increases linearly over a wide range of fields, until fifth order terms start to contribute.
- $\beta = \infty$ (non-polar molecules: $\mu = 0$)

$$<\mu_E> = \left[\bar{\alpha}+\frac{2(\Delta\alpha)^2}{45kT}E^2+\dots\right]E$$

This is the ultimate case of antisaturation.

SUMMARY

The considerations show that the third order term in E of the Langevin expansion of $<\mu_E>$ has a saturating effect if it originates mainly from permanent dipole moments, and an enhancing effect (anti-saturation) if it originates from the polarizability anisotropy.

V. THEORIES RELATING THE MOLECULAR QUANTITIES TO THE MACROSCOPIC ONES

Basic studies of the dielectric properties of matter, when not directly related to electrical insulation problems, are aimed at the elaboration of models able satisfactorily to relate the molecular parameters, such as atomic and molecular polarizabilities and respective concentrations, to the macroscopic, experimental quantities which can be measured, such as the dielectric permittivity and its frequency dependence. Considerable progress has already been made along these lines, in the understanding of condensed matter. We shall not discuss this in detail, but will present only a brief outline of the early models which have paved the way to quantitative dielectric spectroscopy.

V.1. Dilute phases

The simplest relation between polarizability and dielectric permittivity is obtained as above by assuming that the field acting locally on the molecular sites is the same as the applied field. This assumption is valid in dilute phases, and gives:

$$\underbrace{\vec{P} = (\varepsilon - \varepsilon_o)\vec{E}}_{\text{(macroscopic)}} = \underbrace{N\alpha\vec{E}}_{\text{(microscopic)}} \qquad \text{(V.1)}$$

or

$$\varepsilon = \varepsilon_o + N\alpha \qquad \text{(V.2)}$$

Let us now return to the macroscopic definition of the polarization given by eqn.(8') of Table 2: $\vec{P} = (\varepsilon - \varepsilon_o)\vec{E}$. As discussed earlier, $\vec{P}$ has three components:

1. The electronic polarization $\vec{P}_e$, which was discussed from the molecular standpoint in sections (III.1) to (III.3). This arises from the valence electrons, and occurs within about 10^{-14} s of application of an ideally sharp step function of field.
2. The ionic polarization $\vec{P}_i$, discussed in section (III.4). This is due to the elastic deformations inside polar molecules, and occurs within about 10^{-12} s of the field being applied.
3. The orientational polarization $\vec{P}_{or}$, discussed in Chapter IV. This is quite slow compared with the other types, and may require as much as 10^{-6} s to reach a steady state.

The last component $\vec{P}_{or}$ is of special interest, since it is directly ✓ related to the magnitude of the permanent dipole moment (cf. eqn. (IV.8)).

In order to obtain $\vec{P}_{or}$, and from it the magnitude of μ , we now anticipate Chapter VII, and consider the permittivity as measured with an a.c. field, at two distinct frequencies, on either side of the dipole relaxation dispersion range:

(a) at a frequency lower than the dipole relaxation frequency (which is typically between 10^3 and 10^6 Hz), there is only a static permittivity ✓ ε_s, and the magnitude of the polarization is

$$P_s = P_e + P_i + P_{or} = (\varepsilon_s - \varepsilon_o)E \qquad (V.3)$$

(b) at a frequency higher than the dipole relaxation frequency but lower than the ionic resonance frequency (10^9 to 10^{11} Hz), a "high frequen- ✓ cy" permittivity is measured. This is usually called ε_∞ , although the subscript ∞ does not mean "infinite" ($\varepsilon \to \varepsilon_o$ if the frequency actually tends to infinity). It simply implies a frequency which is higher than the relaxation frequency. Since the electric field now oscillates at a frequency above the relaxation frequency, the compo- ✓ nent $\vec{P}_{or}$ no longer contributes to the permittivity, and the magni- tude of the polarization is

$$\cdot \, \varepsilon \to \varepsilon_o$$
$$\equiv \chi \to 0$$
$$(\because \varsigma = \varepsilon_o + \chi)$$
$$(P = \chi E)$$

$$P_\infty = P_e + P_i = (\varepsilon_\infty - \varepsilon_o)E \qquad (V.4)$$

with the same remark on the subscript ∞ of P_∞ .

Subtracting (V.4) from (V.3):

$$P_{or} = (\varepsilon_s - \varepsilon_\infty)E \qquad (V.5)$$

and, from eqn.(IV.8),

$$\frac{N\mu^2}{3kT} = \varepsilon_s - \varepsilon_\infty$$

$$\vec{P} = N\alpha\vec{E} \to \vec{P}_{or} = N\alpha_{or}\vec{E} = N\frac{\mu^2}{3kT}\vec{E}$$
$$\therefore N\frac{\mu^2}{3kT}\vec{E} = (\varepsilon_s - \varepsilon_\infty)\vec{E} \qquad (V.6)$$

This relates the microscopic quantity μ to the macroscopic quantities ✓ $\varepsilon_s, \varepsilon_\infty$, N and T .

V.2. Condensed non-polar phases. Lorentz theory

Eqn. (V.6) assumes that the local field acting on the dipoles is the same as the applied field. This is valid only if the dipoles are distant from one other, as in gases or in dilute solutions in non-polar solvents. Otherwise, the local field is higher than the applied field.

The calculation of the local field has been the subject of much work. The local field is normally regarded as the field existing in a cavity inside the material. The field obviously depends on the shape of this cavity, but is always of the form

$$\vec{E}_{loc} = \vec{E} + \gamma\vec{P} \tag{V.7}$$

where γ is called the depolarization factor.
The first calculation of γ is due to Lorentz. According to this model, a given molecule in a condensed phase in an applied field can be treated as sitting alone in the centre of an empty spherical cavity, on the boundary of which a charge is distributed so that the field lines are not affected by the presence of the cavity. In other words, this model assumes that the cavity wall is "buttered" by a charge distribution which cancels the induced dipole moment of the empty cavity.

Although the conditions stated above are not always clearly stated, this model is treated in all the textbooks, and it is elementary, using the boundary condition on the electric displacement at the interface of this artificial cavity, to find the uniform field inside:

$$E_{loc} = E + \frac{P}{3\varepsilon_o} \tag{V.8}$$

where P is the polarization outside the cavity.

From the definition of P ,

$$E_{loc} = E + \frac{\varepsilon - \varepsilon_o}{3\varepsilon_o} E = \frac{\varepsilon + 2\varepsilon_o}{3\varepsilon_o} E \tag{V.9}$$

solution, as before, gives:

$(\varepsilon-\varepsilon_0)E = P = N p_{loc} = N\alpha E_{loc}$

$$(\varepsilon - \varepsilon_o)E = N\alpha E_{loc} = N\alpha \frac{\varepsilon + 2\varepsilon_o}{3\varepsilon_o} E \tag{V.10}$$

(macroscopic) (microscopic)

or

$$\boxed{\frac{\varepsilon - \varepsilon_o}{\varepsilon + 2\varepsilon_o} = \frac{N\alpha}{3\varepsilon_o}} \tag{V.11}$$

which is the well-known Clausius-Mossotti-Lorentz (CML) relation.

Obviously, this relation is wrong if $\frac{N\alpha}{3\varepsilon_o}$ approaches unity, since this implies $\varepsilon \rightarrow \infty$. It can only be regarded as a fair approximation in the case of a condensed non-polar material (α small) or in the case of a dilute solution of polar molecules in vacuo or in a non-polar solvent.

With this restriction, the CML relation can be used at low or high frequencies, i.e.

at low frequencies (dipole orientation contributes to ε):

$$\frac{\varepsilon_s - \varepsilon_o}{\varepsilon_s + 2\varepsilon_o} = \frac{N}{3\varepsilon_o}(\alpha + \alpha_{or}) \qquad (V.12)$$

the molecular polarizability is both electronic, ionic (α) and orientational (α_{or})

at high frequencies (dipole orientation does not contribute):

$$\frac{\varepsilon_\infty - \varepsilon_o}{\varepsilon_\infty + 2\varepsilon_o} = \frac{N}{3\varepsilon_o}\alpha \qquad (V.13)$$

The polarizability is only electronic and atomic.
Subtracting (V.13) from (V.12) gives:

$$\frac{\varepsilon_s - \varepsilon_o}{\varepsilon_s + 2\varepsilon_o} - \frac{\varepsilon_\infty - \varepsilon_o}{\varepsilon_\infty + 2\varepsilon_o} = \frac{N\alpha_{or}}{3\varepsilon_o} = \frac{N\mu^2}{9\varepsilon_o kT}$$

or

$$\frac{9\varepsilon_o^2(\varepsilon_s - \varepsilon_\infty)}{(\varepsilon_s + 2\varepsilon_o)(\varepsilon_\infty + 2\varepsilon_o)} = \frac{N\mu^2}{3kT} \qquad (V.14)$$

Finally, the difference $(\varepsilon_s - \varepsilon_\infty)$ is:

$$\varepsilon_s - \varepsilon_\infty = \frac{(\varepsilon_s + 2\varepsilon_\infty)(\varepsilon_\infty + 2\varepsilon_o)}{9\varepsilon_o^2}\frac{N\mu^2}{3kT} \qquad (V.15)$$

This equation modifies eqn.(V.6) by the numerical factor

$$A = \frac{(\varepsilon_s + 2\varepsilon_o)(\varepsilon_\infty + 2\varepsilon_o)}{9\varepsilon_o^2} \qquad (V.16)$$

to account for the local field. As expected, the factor A tends to unity if $\varepsilon_s \simeq \varepsilon_\infty \simeq \varepsilon_o$ (dilute phase). Eqn. (V.15) constitutes a significant improvement over eqn. (V.6). Its limits are those of the CML relation, i.e. a divergence of ε_s if $N\alpha \to 3\varepsilon_o$.

A more refined treatment accounting for the presence of permanent dipoles has been given by L. Onsager.

V.3. Condensed phases. Onsager theory

In Onsager's model, the molecule is a polarizable point dipole in the centre of a real cavity of molecular dimensions. Further away than a molecular distance from the centre of the molecule, the material is regarded as a linear continuum of isotropic permittivity ε, submitted to the influence of the molecule, in addition to the applied field. ε depends on the frequency, since the contribution from dipole orientation decreases as the frequency increases.

The various steps in the derivation are as follows :

The polarizable dipole is first considered in the cavity without an external field ; the real cavity with no dipole inside is then considered in the applied field. Finally, the principle of superposition is used to describe the cavity with the dipole inside, in the applied field.

These main steps, together with the relevant definitions, are illustrated in Table 6. However, for complete clarity, the reaction field ($\vec{E}_r$) and the internal field ($\vec{E}_j$) are defined here.

Reaction field $\vec{E}_r$. If a point dipole is sitting in the centre of a real cavity of radius R, it produces a field which polarizes the cavity, distributing on its surface a non-uniform charge density. This charge density in turn creates a uniform "reaction field" $\vec{E}_r$ which is parallel to the dipole.

Problem.

Show, by solving Laplace's equation in the cavity that the reaction field is given by:

$$\vec{E}_r = \frac{2}{4\pi\varepsilon_0 R^3} \cdot \frac{\varepsilon - \varepsilon_0}{2\varepsilon + \varepsilon_0} \vec{\mu}'$$

where $\vec{\mu}' = \vec{\mu} + \alpha \vec{E}_r$ is the effective dipole.

Internal field. Whereas $\vec{E}$ is the uniform field far away from the empty cavity, the internal field $\vec{E}_i$ is the uniform field inside the empty cavity. Again, $\vec{E}_i$ can be derived from Laplace's equation, with even simpler boundary conditions. $\vec{E}_i$ is parallel to $\vec{E}$, and its magnitude turns out to be larger by a factor $3\varepsilon(2\varepsilon + \varepsilon_0)^{-1}$, because of the enhancing effect of the polarization charges at the boundary.

After a clear and complete understanding of Table 6 has been acquired, it becomes easy to relate the molecular properties (N, α, μ, ...) to the macroscopic permittivity, as before, by calculating the Boltzmann average of the projection of the effective dipole moment μ^*, in the direction of the applied field. Here, this projection is:

$$\mu^*_\theta = \mu' \cos\theta + m \tag{V.17}$$

where μ' and m are defined in Table 6.

In order to calculate the Boltzmann average of μ^*_θ, this quantity must be weighted by the appropriate Boltzmann factor $\exp(\frac{-W}{kT})$, where W is the orientational potential energy. However, the second term m of μ^*_θ corresponds to the component of μ^* which remains parallel to the applied field (column 1), independent of the orientation of $\vec{\mu}$. This term should not appear in W, which reduces to

$$W = - \vec{\mu'}.\vec{E_i} = - g\ \mu' E \cos\theta \tag{V.18}$$

Hence, the Boltzmann average of μ^*_θ becomes:

$$<\mu^*_\theta> = \int (\mu' \cos\theta + m)\exp(g\ \frac{\mu' E}{kT} \cos\theta)\,d\Omega \Big/ \int \exp(g\ \frac{\mu' E}{kT} \cos\theta)\,d\Omega$$

where $d\Omega$ is $2\pi \sin\theta\, d\theta$, as in chapter IV.

In order to simplify the notation, we use the following substitutions:

$$x = \cos\theta, \text{ as before,}$$

$$\alpha' = \frac{\alpha\ g}{1 - \alpha f}$$

and

$$y' = g\ \frac{\mu' E}{kT}\ , \quad \text{with} \quad \mu' = \frac{\mu}{1 - \alpha f}$$

(Cf.Table 6)

Thus

$$<\mu^*_\theta> = \int_{-1}^{+1} (\mu' x + \alpha' E)\ \exp(y'x)\,dx \Big/ \int_{-1}^{+1} \exp(y'x)\,dx \tag{V.19}$$

Since $y' \ll 1$, the exponentials can be approximated by an expansion including first order terms only, so that

$$<\mu^*_\theta> \simeq \int_{-1}^{+1} (\mu' x + \alpha' E)\ (1 + y'x)\,dx \Big/ \int_{-1}^{+1} (1 + y'x)\,dx \tag{V.20}$$

$$<\mu^*_\theta> \simeq \frac{1}{3}\ \mu'\ y' + \alpha' E \tag{V.21}$$

Substituting the original variables, eqn. (V.21) becomes

TABLE 6 Onsager's model

DIPOLE MOMENTS

Molecule in cavity, E = 0

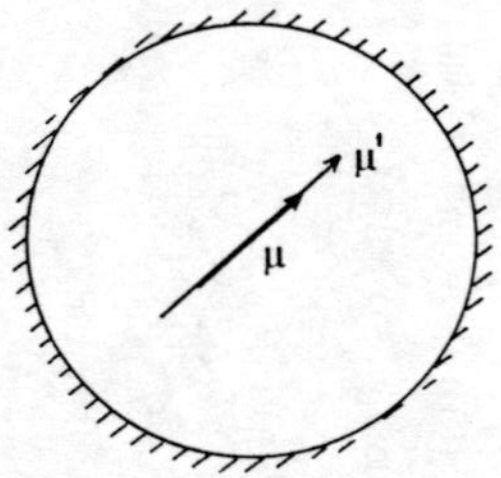

$$\mu' = \mu + \alpha E_r$$

$$E_r = \frac{2}{4\pi\varepsilon_o R^3} \frac{\varepsilon - \varepsilon_o}{2\varepsilon + \varepsilon_o} \mu' = f\mu' \quad \oplus$$

$$\mu' = \mu + \alpha f \mu' \Longrightarrow \mu' = \frac{\mu}{1 - \alpha f}$$

Empty cavity, E ≠ 0

<u>Definitions and symbols</u>

E_r : Reaction field
μ : Actual permanent dipole
μ' : Effective permanent dipole
$\oplus$: From electrostatics
E_i : Internal field
m : Effective induced dipole in cavity
μ^* : Resultant effective dipole in cavity (permanent + induced)

$$\mu^*_\theta = \text{Proj}\,(\mu^*) = \mu' \cos\theta + m$$

Molecule in cavity, E ≠ 0

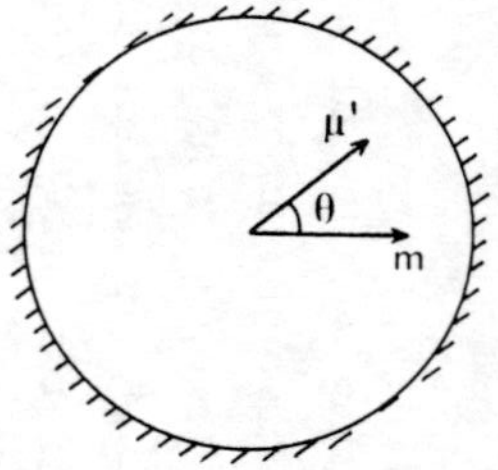

$$m = \alpha\,(E_i + fm)$$

$$m = \frac{\alpha g}{1 - \alpha f} E$$

ELECTRIC FIELDS

Molecule in cavity, E = 0

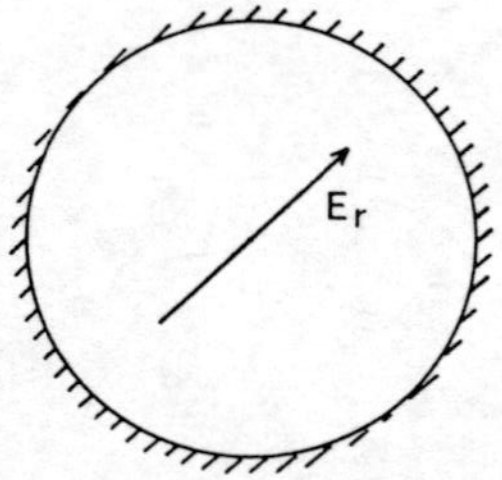

Reaction field:

$$E_r = f\mu' = \frac{f}{1 - \alpha f} \mu$$

Empty cavity, E ≠ 0

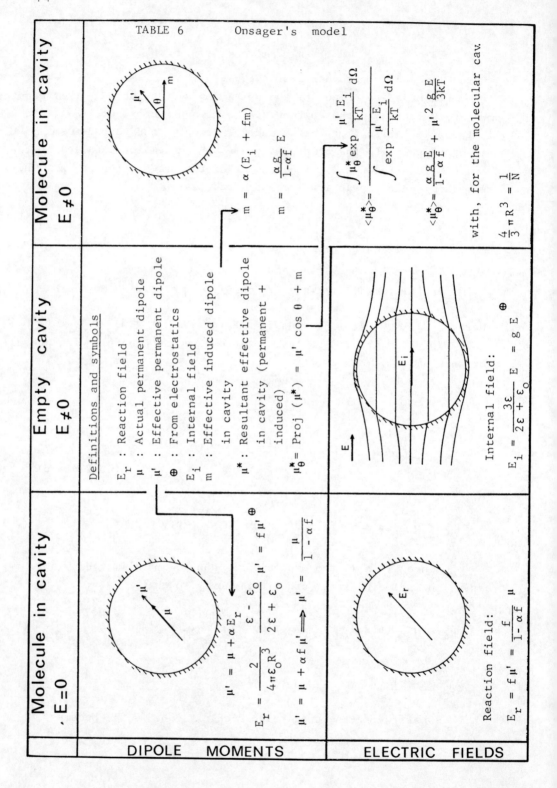

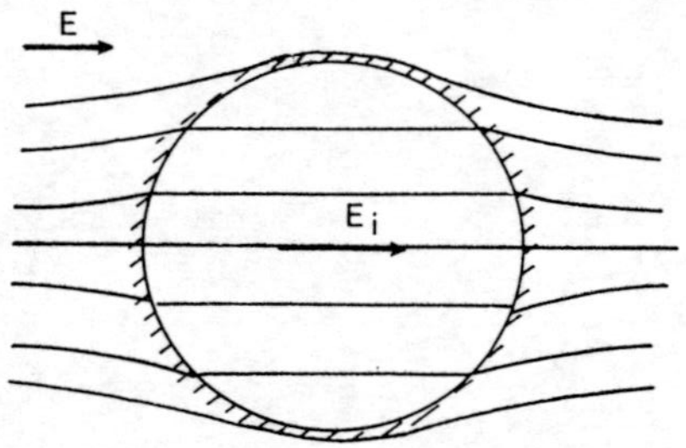

Internal field:

$$E_i = \frac{3\varepsilon}{2\varepsilon + \varepsilon_o} E = g E \quad \oplus$$

Molecule in cavity, E ≠ 0

$$\langle\mu^*_\theta\rangle = \frac{\int \mu^*_\theta \exp \frac{\mu' . E_i}{kT} d\Omega}{\int \exp \frac{\mu' . E_i}{kT} d\Omega}$$

$$\langle\mu^*_\theta\rangle = \frac{\alpha g E}{1 - \alpha f} + \mu'^2 \frac{g E}{3kT}$$

with, for the molecular cav.

$$\frac{4}{3} \pi R^3 = \frac{1}{N}$$

$$\langle\mu_\theta^*\rangle \simeq \frac{g}{1-\alpha f}\left[\alpha + \frac{\mu^2}{3(1-\alpha f)kT}\right]E \tag{V.22}$$

From the two definitions of the polarization:

$$P = (\varepsilon-\varepsilon_o)E = N\langle\mu_\theta^*\rangle,$$

we obtain

$$\varepsilon - \varepsilon_o = \frac{Ng}{1-\alpha f}\left[\alpha + \frac{\mu^2}{3(1-\alpha f)kT}\right] \tag{V.23}$$

This equation is valid for condensed phases, provided that the magnitude of the permanent dipoles is such that the electrostatic interactions between neighbouring dipoles are negligible.

If spherically symmetrical molecules are packed together to form a condensed phase, the concentration N is just the reciprocal of the molecular volume, so that

$$\frac{4}{3}\pi NR^3 = 1 \tag{V.24}$$

and thus

$$f = \frac{2N}{3\varepsilon_o}\,\frac{\varepsilon-\varepsilon_o}{2\varepsilon+\varepsilon_o} \tag{V.25}$$

Equations (V.23) and (V.25) have important consequences when the frequency is very high or very low with respect to the relaxation frequency.

1. Very high frequency

The dipoles do not contribute to the permittivity ε_∞ and equations (V.23) (V.25) become:

$$\varepsilon_\infty - \varepsilon_o = \frac{N\alpha g_\infty}{1-\alpha f_\infty} \tag{V.26}$$

where

$$g_\infty = \frac{3\varepsilon_\infty}{2\varepsilon_\infty+\varepsilon_o}$$

and

$$f_\infty = \frac{2N}{3\varepsilon_o}\,\frac{\varepsilon_\infty-\varepsilon_o}{2\varepsilon_\infty+\varepsilon_o}$$

Introducing these values of f_∞ and g_∞ into eqn.(V.26) gives:

$$\varepsilon_\infty - \varepsilon_o = (\varepsilon_\infty + 2\varepsilon_o)\frac{N\alpha}{3\varepsilon_o}, \tag{V.27}$$

This is the CML relation, obtained as a particular case of Onsager's theory.

2. Very low frequency

In this case, the subscript s is used, and f and g become:

$$g_s = \frac{3\varepsilon_s}{2\varepsilon_s + \varepsilon_o} \tag{V.28}$$

and

$$f_s = \frac{2N}{3\varepsilon_o} \frac{\varepsilon_s - \varepsilon_o}{2\varepsilon_s + \varepsilon_o}$$

Introducing $\frac{N}{3\varepsilon_o}$ as given by the CML relation into this expression for f_s gives:

$$f_s = \frac{2(\varepsilon_\infty - \varepsilon_o)(\varepsilon_s - \varepsilon_o)}{\alpha(\varepsilon_\infty + 2\varepsilon_o)(2\varepsilon_s + \varepsilon_o)}$$

The important factor $(1-\alpha f_s)$ becomes

$$1-\alpha f_s = \frac{3\varepsilon_o}{(\varepsilon_\infty + 2\varepsilon_o)} \frac{(2\varepsilon_s + \varepsilon_\infty)}{(2\varepsilon_s + \varepsilon_o)} \tag{V.29}$$

and by using the CML relation, eqn.(V.29) becomes

$$1-\alpha f_s = \frac{N\alpha}{\varepsilon_\infty - \varepsilon_o} \frac{2\varepsilon_s + \varepsilon_\infty}{2\varepsilon_s + \varepsilon_o} \tag{V.30}$$

A low frequency form of eqn.(V.23) may be written, using eqn. (V.28) for g_s and eqn.(V.30) for $(1-\alpha f_s)$. Rearrangement gives:

$$\boxed{\varepsilon_s - \varepsilon_\infty = \frac{\varepsilon_s (\varepsilon_\infty + 2\varepsilon_o)^2}{3\varepsilon_o^2 (2\varepsilon_s + \varepsilon_\infty)} \frac{N\mu^2}{3kT}} \tag{V.31}$$

This is Onsager's relation; it is similar to eqn.(V.6), with the new corrective factor

$$A' = \frac{\varepsilon_s (\varepsilon_\infty + 2\varepsilon_o)^2}{3\varepsilon_o^2 (2\varepsilon_s + \varepsilon_\infty)} = \frac{K_s(K_\infty + 2)^2}{3(2K_s + K_\infty)} \tag{V.32}$$

The factor A' may be significantly different from the factor A which was obtained by direct use of the CML relation. If K_s is not large compared to unity, A and A' are very close to each other, and of the order of unity. If K_s is very large compared to unity, as in the case of water, for instance, the old CML corrective factor varies as $\frac{K_s}{9}(K_\infty+2)$, but the Onsager factor remains of the order of $\frac{1}{6}(K_\infty+2)^2$, which is more satisfactory.

This does not mean that Onsager's model is adequate to interpret the dielectric properties of special substances like water, in which hydrogen bonds behave as very unusual dipoles, but at least it is far superior to models based on the CML relation.

Onsager's theory itself has been and still is being improved. The most obvious improvement, suggested by Fröhlich, is to take into account, in Onsager's model, the specific directional interaction which may exist between the molecules.

If γ_{ij} is the angle between the dipoles i and j, the average of $\cos\gamma_{ij}$ over all pairs in an isotropic material is zero, if the average is taken over a radius large compared to the molecular dimensions.

There are cases, however, where a selective orientational interaction producing a local anisotropy prevents the average of $\cos\gamma_{ij}$ from vanishing. This occurs particularly for many materials containing hydrogen bonds. Cyanhydric acid HCN is a particularly relevant case; the molecules align themselves in a polymer-like structure (a "labile" polymer), and $\langle\cos\gamma_{ij}\rangle$ is close to unity.

In such a case, one can define, for the molecule i, an orientational correlation parameter

$$g_i = 1 + \sum_j \cos\gamma_{ij}$$

where the summation is carried over the interacting neighbours. The local field, and hence the corrective factor of Onsager's theory, should be multiplied by a factor g larger than unity.

Calculations have been carried out to include in Onsager's theory the non-spherical symmetry and, in particular, the polarizability anisotropy of many molecules. They have been applied, with notable success, to liquid crytals, which are made of long, rigid molecules. Theses calculations lie outside the scope of this book, but reference to sections (III.3) and (V.3) should make them easily accessible.

In Table 7, the corrections resulting from the various local field models discussed above ($E_{loc} = E$, CML, Onsager and Fröhlich models) are compared.

TABLE 7 - Comparison of the local field corrections for various models.

Model	Corrective factor A in $\varepsilon_s - \varepsilon_\infty = A \frac{N\mu^2}{3kT}$
Very dilute phase ($E_{loc} = E$)	1
Clausius - Mossotti - Lorentz	$\frac{(K_s + 2)(K_\infty + 2)}{9}$
Onsager	$\frac{K_s (K_\infty + 2)^2}{3(2K_s + K_\infty)}$
Fröhlich	$g \frac{K_s (K_\infty + 2)^2}{3(2K_s + K_\infty)}$

V.4. The Kerr electro-optic effect

1. Definition

The Kerr effect is the birefringence induced in a fluid by a strong electric field. To a first approximation, this birefringence is a quadratic function of the field E, so that the Kerr constant B of a fluid can be defined by

$$\Delta n = B E^2 \qquad (V.33)$$

where Δn is the difference between the refractive index $n_{/\!/}$ for a polarized light wave of wavelength λ and electric vector parallel to the applied field (see Table 8), and the refractive index $n_\perp$ for the same polarized light wave with its electric vector perpendicular to the applied field. Here, the dimensions of B are $[L^2 V^{-2}]$. B is usually defined, however, from the equation $\Delta n = \lambda B E^2$, where λ is the wave length of the light. In this case, the dimensions of B are $[L V^{-2}]$.

The permanent dipoles, being very sluggish with respect to the period of the light wave, which is of the order of 10^{-14}sec, do not contribute directly to the refractive index, but only through their action on the orientation of the molecules. In other words, the refractive index corresponds to the electronic polarizability of the molecule oriented by the applied field.

TABLE 8 - Schematic description of the electro-optic Kerr effect.

CONFIGURATION $\vec{e}$ = electric vector of incoming light.	ORIENTATION of $\vec{e}$ with respect to the applied field $\vec{E}$.	Refractive index
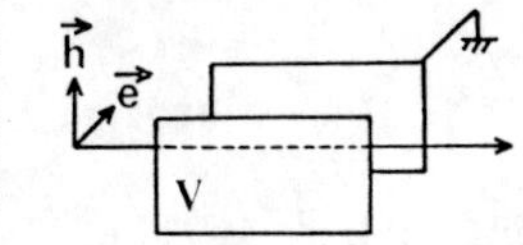	$\vec{e}$ // $\vec{E}$	$n_{//}(E)$ usually $>$ n(0)
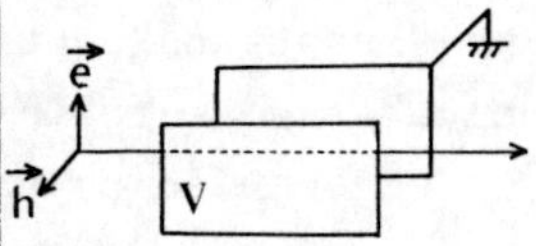	$\vec{e} \perp \vec{E}$	$n_{\perp}(E)$ usually $<$ n(0)
K E R R E F F E C T		$n_{//} - n_{\perp} = B E^2$

2. Elementary derivation of the Kerr constant

In order to calculate B , we must calculate the refractive index (i.e. the square root of the high frequency permittivity) for the two configurations of Table 8 .

The high frequency permittivity $\varepsilon_{//}$ obtained with the measuring (optical) field parallel to the applied static field E may be written

$$\varepsilon_{//} = \varepsilon_0 + N \frac{\partial}{\partial E} \langle\mu_E\rangle \tag{V.34}$$

Curly "∂" are used here since the derivation is carried out for constant θ (cf. Chapter IV). The electric vector of the light constitutes a very small, very high frequency modulation of the applied d.c. field, and has no orienting effect on the molecules. Therefore, using eqn.(IV.22) for $\langle\mu_E\rangle$, $\varepsilon_{//}$ becomes:

$$\varepsilon_{//} = \varepsilon_0 + N\left[\Delta\alpha \langle\cos^2\theta\rangle + \alpha_2\right] \tag{V.35}$$

Using the value found previously (section (IV.3)) for $\langle\cos^2\theta\rangle$, one obtains:

$$\varepsilon_{/\!/} = \varepsilon_0 + \frac{N}{3}\bar{\alpha} + \frac{2N\Delta\alpha}{45kT}\left(\Delta\alpha + \frac{\mu^2}{kT}\right)E^2 + \ldots \qquad (V.36)$$

so that

$$n_{/\!/} = \left(\frac{\varepsilon}{\varepsilon_0}\right)^{1/2} = \left[1 + \frac{N\bar{\alpha}}{3\varepsilon_0} + \frac{2N\Delta\alpha}{45kT\varepsilon_0}\left(\Delta\alpha + \frac{\mu^2}{kT}\right)E^2 + \ldots\right]^{1/2}$$

If the concentration N is small enough so that $\frac{N\bar{\alpha}}{3\varepsilon_0} \ll 1$,

$$n_{/\!/} \simeq 1 + \frac{N\bar{\alpha}}{6\varepsilon_0} + \frac{N\Delta\alpha}{45kT\,\varepsilon_0}\left(\Delta\alpha + \frac{\mu^2}{kT}\right)E^2 \qquad (V.37)$$

The first two terms represent the refractive index n(0) in the absence of any applied field and the last term is due to the field.

The high frequency permittivity $\varepsilon_\perp$ with the "measuring" optical field perpendicular to the applied field can be calculated as follows. From general symmetry considerations of the type which enabled us to define the average polarizability $\bar{\alpha}$ it can be shown that:

$$\varepsilon_{/\!/} + 2\varepsilon_\perp = 3\,\varepsilon(0) \qquad (V.38)$$

Writing $\varepsilon_{/\!/}$ in the form $\varepsilon_{/\!/} = \varepsilon(0) + \Delta\varepsilon$,

$$\varepsilon_\perp = \varepsilon(0) - \frac{\Delta\varepsilon}{2}$$

or

$$\varepsilon_\perp = \varepsilon_0 + \frac{N}{3}\bar{\alpha} - \frac{N\Delta\alpha}{45kT}\left(\Delta\alpha + \frac{\mu^2}{kT}\right)E^2 \ldots$$

and, using the same approximation as above:

$$n_\perp = \left(\frac{\varepsilon_\perp}{\varepsilon_0}\right)^{1/2} \simeq 1 + \frac{N\bar{\alpha}}{6\varepsilon_0} - \frac{N\Delta\alpha}{90kT\varepsilon_0}\left(\Delta\alpha + \frac{\mu^2}{kT}\right)E^2 \qquad (V.39)$$

From (V.37) and (V.39),

$$\Delta n = \frac{N\Delta\alpha}{30kT\varepsilon_0}\left(\Delta\alpha + \frac{\mu^2}{kT}\right)E^2$$

so that the birefringence coefficient, as defined here, is :

$$\boxed{B = \frac{N\Delta\alpha}{30kT\varepsilon_0}\left(\Delta\alpha + \frac{\mu^2}{kT}\right)} \qquad (V.40)$$

This approximate equation is extremely important, as it shows that the Kerr constant of a fluid is given essentially by its polarizability anisotropy. If the polarizability is isotropic, the equation shows that the fluid does not exhibit Kerr birefringence, no matter how large is the permanent dipole moment of the molecule. Conversely, provided that $\Delta\alpha$ is positive, the Kerr constant exists, and increases with μ^2. Remembering previous results about induced and permanent dipoles, in most polar molecules at room temperature, $\frac{\mu^2}{kT} \gg \Delta\alpha$, so that an even simpler approximation can be obtained by neglecting $\Delta\alpha$ in comparison with $\frac{\mu^2}{kT}$ in eqn.(V.40).

Equation (V.40) shows that, within the frame of the well defined approximations used, it is a necessary and sufficient condition for the existence of electrical birefringence in a gas or a solution in a non-polar solvent, that the molecules of solute have a strong polarizability anisotropy $\Delta\alpha$. In other words, the molecules likely to yield a high Kerr constant are elongated molecules of high average polarisability.

The presence of a permanent dipole μ contributes to, but does not produce, by itself, electrical birefringence.

As an illustration, the existence in the molecules of one or several aromatic rings, with highly polarizable delocalized π electrons, favours Kerr effect. Hence, if we simply replace a hydrogen atom in benzene by a halogen (for instance chlorine, giving chlorobenzene) or by a polar group such as - N = O , giving nitrobenzene, or - C ≡ N, giving cyanobenzene, etc., we expect to obtain a molecule with a high Kerr constant. This is the case, as is shown in the table 9 below, which gives the Kerr constant $B' = \frac{\Delta n}{\lambda E^2}$ of most of the common liquids, as quoted in the Londolt-Börnstein tables (Bd II/8, pages 849 to 870).

Some other molecules, particularly alcohols, display a negative Kerr constant. This may occur when the permanent dipole moment $\vec{\mu}$ makes an angle $\Psi \neq 0$ with the principal molecular axis z (Cf. the problem and Table 10).

Problem

Using the basic equation of spherical trigonometry, carry out the averaging procedure on $\cos^2\theta$, taking into account the angle $\Psi = (z, \vec{\mu})$ and the new variable φ , the angle between the planes $(z, \vec{\mu})$ and $(z, \vec{E})$ where z is the long molecular axis.

Show that in eqn. (V.27) for $\langle\cos^2\theta\rangle$, μ^2 should now be multiplied by $(1 - \frac{3}{2}\sin^2\Psi)$, so that the Kerr constant is now:

$$B = \frac{N\Delta\alpha}{30kT}\left[\Delta\alpha + \frac{\mu^2}{kT}\left(1 - \frac{3}{2}\sin^2\Psi\right)\right] \tag{V.41}$$

This coefficient is negative if Ψ is close enough to $\pi/2$, and $z = \frac{2\,\Delta\alpha kT}{\mu^2}$ smaller than unity.

TABLE 9 - Positive Kerr constants of some common pure liquids.

Definition $B' = \Delta n/E^2\lambda$ [$L\ V^{-2}$]

(from the Landolt-Börnstein tables, 6 Bd II/8 pages 849-870)

NAME	FORMULA	$10^9 B'$ (e.s)	$10^{16} B'$ (S.I)	Experimental conditions
methane	CH_4	2		172 K
n-pentane	C_5H_{12}	5.5	6.1	R.T.
cyclohexane	C_6H_{12}	5.9	6.55	-
n-hexane	C_6H_{14}	6.6	7.33	-
carbon tetrachloride	CCl_4	7.1	7.88	-
nitrogen	N_2	8.0	8.88	77.4 K
n-decane	$C_{10}H_{22}$	10.3	11.4	R.T.
carbon dioxide	CO_2	14	15.5	-
ethylene	C_2H_4	18	20	169 K
oxygen	O_2	20	22.2	90 K
benzene	C_6H_6	41	45.5	R.T.
ethyl alcohol	C_2H_5OH	85	94.4	R.T.
p-dichlorobenzene	$C_6H_4Cl_2$	263	292	-
carbon disulphide	CS_2	360	400	-
m-dichlorobenzene	$C_6H_4Cl_2$	890	988	-
chlorobenzene	C_6H_5Cl	1,050	1,067	-
nitromethane	CH_3NO_2	1,200	1,330	-
acetone	CH_3COCH_3	1,800	2,000	-
pyridine	C_6H_5N	2,275	2,527	-
o-dichlorobenzene	$C_6H_4Cl_2$	4,400	4,880	-
o-nitrotoluene	$C_6H_4CH_3NO_2$	20,000	22,200	-
m-nitrotoluene	$C_6H_4CH_3NO_2$	23,000	25,550	327 K
nitrobenzene	$C_6H_5NO_2$	40,000	44,400	-

This table shows that, unlike the theoretical predictions, some molecules with spherical symmetry (CH_4, CCl_4) have a non-zero Kerr constant. This probably originates from the ionic polarizability of these molecules, which loose their spherical symmetry in a strong applied field.

TABLE 10 - Liquids with negative Kerr constants.

NAME	FORMULA	10^9B (e.s.)	$10^{16}B$ (S.I.)	Experimental conditions
diethylether	$(C_2H_5)_2O$	- 66	- 73.3	R.T.
dichloromethane	CH_2Cl_2	-129	-143	-
chloroform	$CHCl_3$	-300	-333	-
n-butylalcohol	C_4H_9OH	-400	-444	-
n-octylalcohol	$C_8H_{17}OH$	-850	-944	-
cyclohexanol	$C_6H_{11}OH$	-1.030	-1.144	-

3. Applications of the Kerr effect

The Kerr effect for simple molecules has been described here in some detail, because it is a typically rewarding outcome of the overlap between materials science and engineering. The most famous applications for this effect are undoubtedly the light modulator and the light switch, which are shown schematically in Figure 21.

In the absence of an applied field, the polarization of the light remains perpendicular to the analyzer, so that no light goes through the switch, which is therefore "closed". An applied voltage modifies the polarization of the incoming light, so that some of it may be transmitted by the analyzer. The attractive feature of this switch is its very short time constant, which is due to the very short relaxation time ($\tau < 10^{-10}$ sec.) of the common low viscosity polar liquids. Progress in this field is only limited by the technical problems involved in the production and propagation of high-voltage pulses with very short rise and decay times.

Another important application of the Kerr effect is the analysis of field distributions in a transparent insulation submitted to a high voltage. This problem is of paramount importance for electrical engineers since, owing to the presence of space charges, the local field, somewhere in an insulation under stress, can be much larger than the applied field.

In order to optimize the design of capacitors, cables, etc., the field in the insulation should be as uniform as possible, otherwise breakdown may initiate at highly stressed points, while most of the insulation,under a stress much smaller than the breakdown field, is redundant.

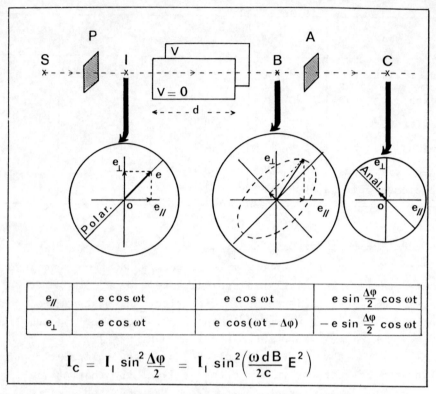

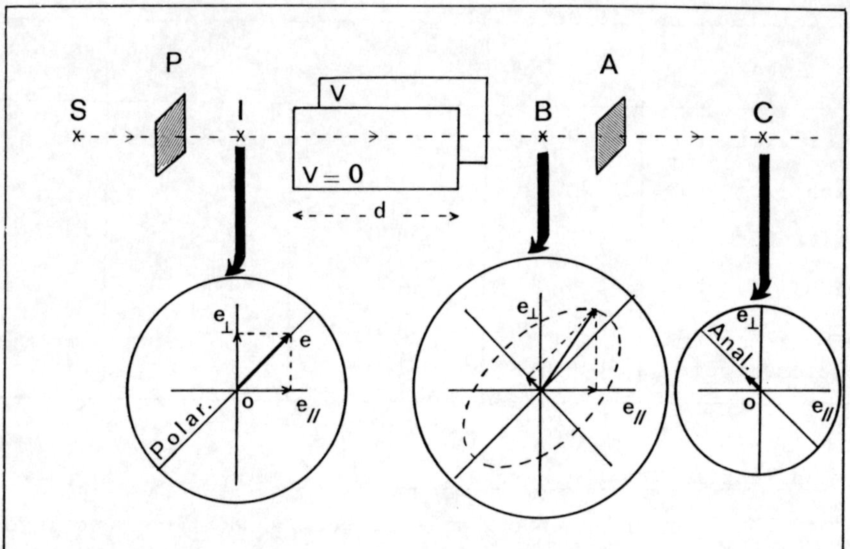

$e_{//}$	$e \cos \omega t$	$e \cos \omega t$	$e \sin \frac{\Delta\varphi}{2} \cos \omega t$
$e_{\perp}$	$e \cos \omega t$	$e \cos(\omega t - \Delta\varphi)$	$- e \sin \frac{\Delta\varphi}{2} \cos \omega t$

$$I_C = I_I \sin^2 \frac{\Delta\varphi}{2} = I_I \sin^2\left(\frac{\omega d B}{2c} E^2\right)$$

Fig. 21 - Principle of the Kerr light switch.
P = Polarizer, A = Analyzer, D = Detector

Several techniques based on the Kerr effect have been developed to analyze the stress distribution in a plane electrode configuration. The simplest one uses the same configuration as that of the light modulator, but with a longer optical path. As above, the polarizer is tilted at 45°, and the analyzer is perpendicular to the polarizer. As can be seen in Figure 21, the incoming electric vector $\vec{e}$ can be resolved into $\vec{e}_{//}$, parallel to the applied field and $\vec{e}_{\perp}$, perpendicular to the applied field, both of magnitude $e\sqrt{2}/2$. These two vectors are originally in phase, but they propagate with their own velocities $v_{//} = \frac{c}{n_{//}}$ and $v_{\perp} = \frac{c}{n_{\perp}}$. After travelling a distance d in the field E , the phase angles of the wave vectors $\vec{e}_{//}$ and $\vec{e}_{\perp}$ are respectively

$$\text{for } \vec{e}_{//} \qquad \varphi_{//} = \frac{\omega d}{c} n_{//}$$

$$\text{for } \vec{e}_{\perp} \qquad \varphi_{\perp} = \frac{\omega d}{c} n_{\perp}$$

with c = velocity of light
and ω = angular frequency of light

Consequently, a phase shift $\Delta\varphi$ appears between the outgoing components $\vec{e}_{/\!/}$ and $\vec{e}_{\perp}$:

$$\Delta\varphi = \frac{\omega d}{c}(n_{/\!/} - n_{\perp}) = \frac{\omega d}{c} BE^2 \qquad (V.42)$$

If the field E is uniform, $\Delta\varphi$ is uniform, and light of uniform intensity is transmitted. If, on the other hand, E is not uniform, the phase angle $\Delta\varphi$ is not uniform, and neither is the amount of light transmitted by the analyzer, which varies in the depth of the sample.

If we assume (Cf Problem and Fig.22) that E varies linearly from 0 (at the electrode on the left) to E_{max} (at the opposite electrode), and, furthermore, that d, B and E are large quantities, there will be extinction of light not only at the left electrode where $E = 0$, but also at each position where $\Delta\varphi$ is a integer multiple of π ,

$$\frac{\omega d}{c} BE^2 = n\pi$$

If the field distribution is $E(x)$, dark fringes parallel to the electrodes are observed at distances x defined by

$$E^2 = \frac{n\pi c}{\omega dB}$$

with $n = 1, 2, 3...$

Problem

Uniform charge distribution.

1. Show that in the case of a uniform charge distribution ρ , the field distribution is of the form

$$E(x) = E_a + \frac{\rho}{\varepsilon}(x - \frac{L}{2})$$

where E_a is the applied field and L the sample thickness. Find the particular value of ρ for which $E(0) = 0$.

2. Under the condition $E(0) = 0$, find the minimum value of distance d which gives only two dark fringes (at $x = 0$ and $x = L$).

3. In general, it can be shown that the mere presence of fringes reveals that the field is not uniform, and that the number of

fringes is a measure of the field distortion.

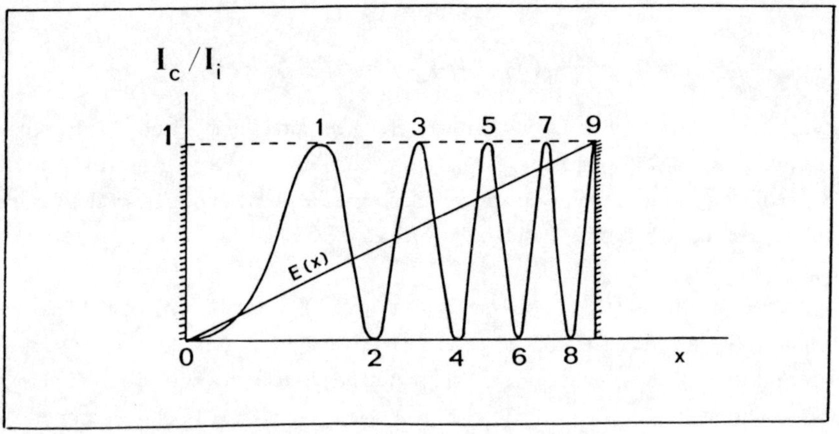

Fig. 22 - Transmitted intensity versus depth for $E = ax$.
Dark fringes at 0, 2, 4, 6, 8

An improvement in the display is obtained by introducing a quarter-wave plate and a rotatory power linear compensator, such as a Babinet double prism between the Kerr cell and the analyzer. The quarter-wave plate transforms the elliptic light emerging from the cell into a linearly polarized wave (Fig.23), and the compensator rotates the polarization of this light by an angle $\theta = ky$. (Fig. 24).

A series of horizontal fringes are now observed, if the field is uniform. The vertical displacement of the fringes when a uniform field is applied is proportional to the square of E . If the field is distorted, the fringes undergo distortions as $E^2(x)$. This elegant method was developed in France, and has been used extensively to analyze the dynamics of the space charges in liquids.

The author of this book is convinced that the Kerr effect is likely to find many more applications in science and in technology. For instance, an elegant method for molecular characterization can be imagined, based on equation (V. 40). From this equation, it is clear that if BT is plotted against $\frac{1}{T}$, one should obtain a straight line of slope $N\mu^2\Delta\alpha/30\ \varepsilon_o k$, which intercepts the BT axis at the point $N(\Delta\alpha)^2/30\varepsilon_o k$. Consequently, with a preset temperature programme for the Kerr cell, B(T) could be derived from signals induced by short pulses of high voltage at regular intervals. Proper processing of the data should yield the main molecular quantities μ and $\Delta\alpha$ in a rather short experimental time, as indicated in Fig.24.

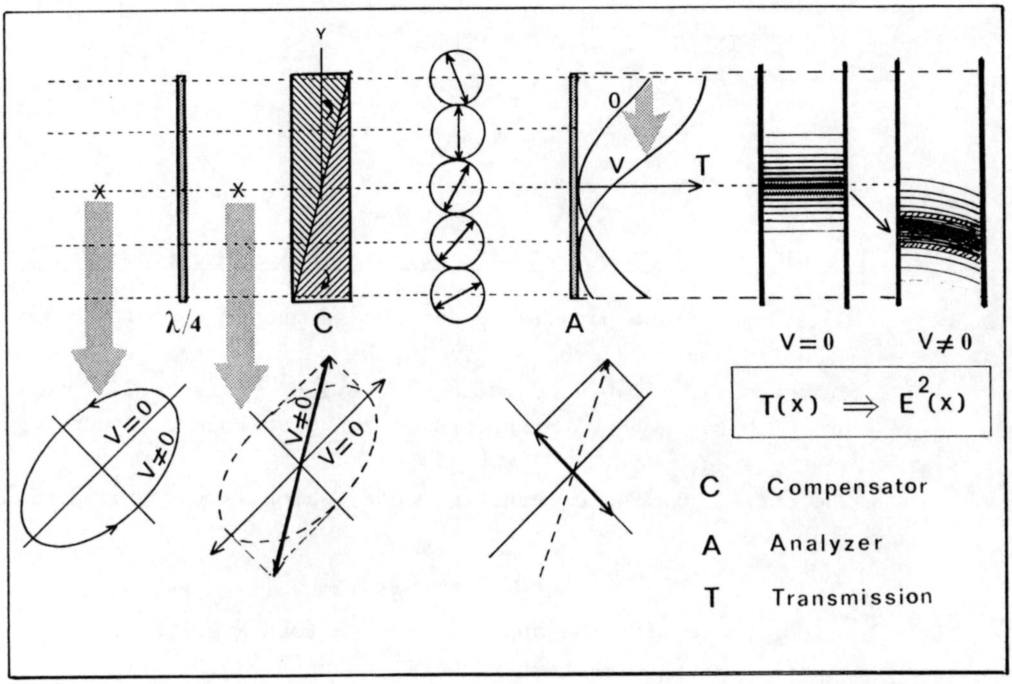

Fig. 23 - Use of a rotatory power compensator.(C)

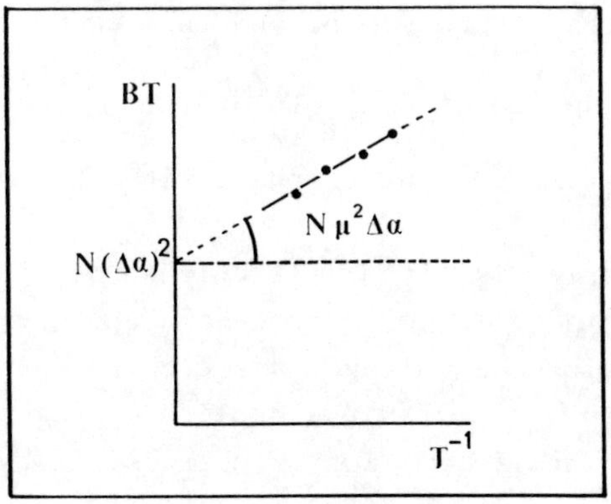

Fig. 24 - Proposed determination of molecular parameters by Kerr constant method.

PART 2. MATTER IN AN ALTERNATING FIELD

VI. THE COMPLEX PERMITTIVITY

VI.1. Definitions of ε^* and σ^*. Propagation of an electromagnetic wave

Consider a plane-parallel condenser of vacuum capacitance $C_o = \varepsilon_o\, S/d$ (S and d are respectively the surface area of the plates and the thickness). If an a.c. voltage $V = V_o e^{i\omega t}$ is applied to this capacitor, a charge $Q = C_oV$ appears on the electrodes, in phase with the applied voltage.

The current in the external circuit, which is the time-derivative of the charge Q:

$$I = \dot{Q} = i\omega\, C_oV \qquad \text{(VI.1)}$$

is 90° out of phase with the applied voltage (cf. Fig.25). It is a non-dissipative displacement or induction current.

We now fill up the volume between the electrodes with a non-polar, perfectly insulating material. The capacitor then displays a new capacitance $C = K\, C_o$ with $K > 1$. The ratio K is the relative permittivity of the sample material, with respect to vacuo, and the new current

$$I' = \dot{Q}' = i\omega\, CV = KI \qquad \text{(VI.2)}$$

is larger than above, but is still 90° out of phase with the applied voltage.

Now, if the sample material is either slightly conducting or polar, or both, the capacitor is no longer perfect and the current is not exactly 90° out of phase with the voltage, since there is a small component of conduction GV in phase with the applied voltage. The origin of this in-phase component is a motion of charges. If these charges are free, the conductance G is effectively independent of the frequency of the applied voltage, but if these charges are bound to opposite charges as in oscillating dipoles, G is a function of frequency, which will be discussed in detail in the chapter dealing with dipole relaxation.

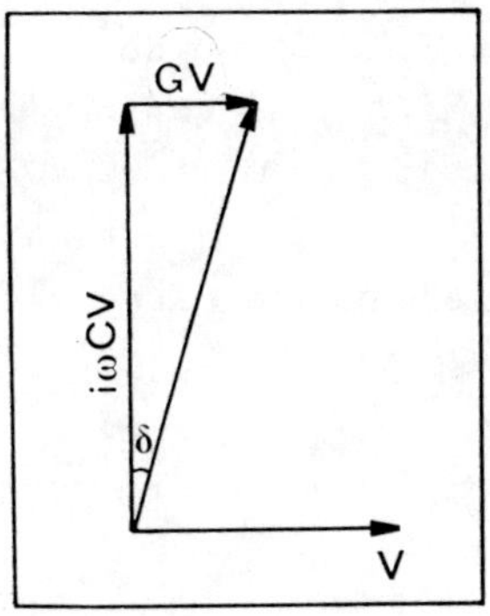

Fig.25 - Current in a leaking capacitor.

In either case, the resultant current is :

$$\vec{I} = (i\omega\, C + G)\, \vec{V} \qquad \text{(VI.3)}$$

If G is pure conductance due to free charges, $G = \sigma S/d$, and since $C = \varepsilon\, S/d$, the density $\vec{j}$ of the current is:

$$\vec{j} = (i\omega\varepsilon + \sigma)\, \vec{E} \qquad \text{(VI.4)}$$

The term $i\omega\varepsilon\,\vec{E}$ is the displacement current density $\dot{\vec{D}}$, and $\sigma\vec{E}$ is the conduction current density.

These may be considered by introducing:

either a complex conductivity σ^* defined by $\vec{j} = \sigma^*\vec{E}$:

$$\sigma^* = i\omega\varepsilon + \sigma \qquad \text{(VI.5)}$$

or a complex permittivity ε^* defined by $\vec{j} = i\omega\,\varepsilon^*\,\vec{E}$:

$$\varepsilon^* = \frac{\sigma^*}{i\omega} = \varepsilon - i\,\frac{\sigma}{\omega} \qquad \text{(VI.6)}$$

The latter quantity will usually be used in the study of dielectrics, since the dissipative term is, in general, very small with respect to the capacitive term.

The loss angle (δ in Fig. 25) is such that

$$\tan\delta = \frac{\text{dissipative term}}{\text{capacitive term}} = \frac{\sigma}{\omega\varepsilon} \qquad \text{(VI.7)}$$

Whenever the conduction (or dissipation) is not exclusively due to free charges, but is also due to bound charges, the conductivity σ is itself a complex quantity which depends on the frequency, so that the real part of ε^* is not exactly ε, and the imaginary part is not exactly $\frac{|\sigma|}{\omega}$.

The most general expression for the complex permittivity is:

$$\varepsilon^* = \varepsilon' - i\varepsilon'' \qquad \text{(VI.8)}$$

where ε' and ε'' are frequency-dependent. These two components and their frequency dependence will be studied after the various types of interactions between the electromagnetic field and matter have been introduced.

The propagation of an electromagnetic wave in matter is based on Maxwell's equations:

$$\vec{\nabla} \times \vec{E} = -\frac{\partial \vec{B}}{\partial t} \quad , \quad \vec{B} = \mu^* \vec{H} \quad (\mu^* = \text{complex magnetic permeability})$$

$$\vec{\nabla} \times \vec{H} = \frac{\partial \vec{D}}{\partial t} \quad , \quad \vec{D} = \varepsilon^* \vec{E} \quad (\varepsilon^* = \text{complex permittivity}).$$

Elimination of $\vec{H}$ gives the propagation equation:

$$\nabla^2 \vec{E} - \varepsilon^* \mu^* \frac{\partial^2 \vec{E}}{\partial t^2} = 0 \qquad \text{(VI.9)}$$

In a cartesian coordinate system where the x axis is chosen along the direction of propagation, eqn. (VI.9) reduces to

$$\frac{\partial^2 E}{\partial x^2} - \varepsilon^* \mu^* \frac{\partial^2 E}{\partial t^2} = 0 \qquad \text{(VI.10)}$$

In other words, electromagnetic waves propagate through a material defined by ε^* and μ^* with a complex velocity $v^* = (\varepsilon^* \mu^*)^{-1/2}$, whereas they propagate in vacuo with a velocity $c = (\varepsilon_o \mu_o)^{-1/2}$. The ratio c/v^* defines the complex index of refraction ; $n^* = n - ik$.

$$n^* = \left(\frac{\varepsilon^* \mu^*}{\varepsilon_o \mu_o} \right)^{1/2} = n - ik \qquad \text{(VI.11)}$$

The general solution of eqn. (VI.10) is of the form:

$$E = E_o \exp(i\omega t - \gamma^* x) \qquad \text{(VI.12)}$$

where $\gamma^* = \alpha + i\beta$ is a complex absorption coefficient. If the value of γ^* is introduced into eqn. (VI.12) this equation becomes:

$$E = E_o e^{-\alpha x} \exp i(\omega t - \beta x) \qquad \text{(VI.13)}$$

In eqn. (VI.13), α appears as the absorption and β as the phase. Furthermore, combining eqns.(VI.10) and (VI.12) gives:

$$\gamma^{*2} + \varepsilon^* \mu^* \omega^2 = \gamma^{*2} + n^{*2} \frac{\omega^2}{c^2} = 0$$

or

$$\gamma^* = i\,\frac{\omega}{c}\,n^* \qquad \text{(V1.14)}$$

Solution for the real and imaginary parts of both members of eqn.(V1.14) gives

$$\begin{pmatrix}\alpha\\ \beta\end{pmatrix} = \frac{\omega}{c}\begin{pmatrix}k\\ n\end{pmatrix}$$

In a non-magnetic material, $\mu^* = \mu_o$, and eqn.(VI.11) reduces to

$$n^* = \left(\frac{\varepsilon^*}{\varepsilon_o}\right)^{1/2} = K^{*\,1/2} \qquad \text{(V1.15)}$$

with $K^* = K' - iK''$ (relative complex permittivity).

This complex relation, which is known as Maxwell's relation, is equivalent to the system:

$$\begin{cases} n^2 - k^2 = K' \\ 2nk = K'' \end{cases}$$

which can be solved for n and k:

$$\begin{cases} n = \dfrac{1}{\sqrt{2}}\left(\sqrt{K'^2 + K''^2} + K'\right)^{1/2} \\[2ex] k = \dfrac{1}{\sqrt{2}}\left(\sqrt{K'^2 + K''^2} - K'\right)^{1/2} \end{cases} \qquad \text{(V1.16)}$$

If the dissipation is small - as is usual in dielectrics away from the critical dispersion frequencies, $K''/K' = \tan\delta \sim \delta$, and

$$\begin{cases} n = \dfrac{c}{\omega}\,\beta \simeq \sqrt{K'} \\[2ex] k = \dfrac{c}{\omega}\,\alpha \simeq \dfrac{\delta}{2}\,\sqrt{K'} = \dfrac{K''}{2\sqrt{K'}} \end{cases} \qquad \text{(V1.17)}$$

From equations (V1.17), the two components of K^* can be expressed in terms of the absorption coefficient α and the phase coefficient β, which are measured by microwave techniques. Finally,

$$K^* = (\beta^2 - 2i\alpha\beta)\left(\frac{c}{\omega}\right)^2 \qquad \text{(V1.18)}$$

Equation (V1.18) constitutes the basis of v.h.f. dielectric spectroscopy.

<u>Problem</u>

Using Maxwell's equations, show that the ratio between the electric and the magnetic vector of the propagating wave is a constant of the material (its "characteristic impedance"), and that the value of this constant is $Z^* = (\mu^*/\varepsilon^*)^{1/2}$.

VI.2. The various types of charges and charge groups, and the corresponding interactions

In any material, there are various types of charges and charge associations, which we now consider.

(a) The "inner" electrons (i.e. those of the inner electronic shells), tightly bound to the nuclei. Although little affected by the applied field, they "resonate" with high energy ($\simeq 10^4$ eV), short wave length ($\simeq 10^{-10}$ m) electromagnetic fields corresponding to the X-ray range.

(b) The "outer" electrons (i.e. those of the outer electronic shells). These are the valence electrons, which contribute to the atomic and/or molecular polarizabilities, and also, in the case of elongated molecular structures, to their orientation with respect to the applied field.

(c) The free electrons or conduction electrons, which contribute to the "in phase" conduction. When submitted to an electric field $\vec{E}$, these electrons move with a velocity $\vec{v} = \mu \vec{E}$. The mobility μ, characteristic of a given material at a given temperature, accounts for all the inelastic collisions which, on average, confer to the electrons a velocity $\mu \vec{E}$ in the field $\vec{E}$. Note that if the electron concentration is not uniform (for instance whenever they accumulate near the electrodes), the "diffusion field" $- D\vec{\nabla} n$ also contributes to the motion.

(d) The bound ions, or ions bound to oppositely charged ions, forming molecular dipoles (e.g. $Cl^- H^+$) or a dipole association in a lattice (e.g. $\boxed{Li}^- Mg^+$ in LiF where $\boxed{Li}^-$ is a lithium vacancy and Mg^+ a substitutional cation. These permanent dipoles experience an orientational torque in a uniform field; in a non-uniform field, a net force also acts, in addition to the torque.

(e) The free ions, as in electrolytes and non-stoichiometric ionic crystals (e.g. excess K^+ in K Cl) which move in the applied field, usually with a low mobility. Ionic dipoles such as $(OH)^-$ show both ionic and dipole characteristics.

(f) Finally, the multipoles, and mainly the quadrupoles (cf. Chapter I), or an antiparallel association of two dipoles, which undergo only a configurational strain in a uniform field, and a slight torque in a divergent field.

Let us assume that an electromagnetic field of frequency f is abruptly applied at an instant which defines the origin of t . Depending on its frequency, this field puts into oscillation one or more types of charges or charge associations among those listed above. Each configuration having its own critical frequency, above which the interaction with the field becomes vanishingly small, the lower the frequency, the more configurations are excited. Of course the critical frequency of a given configuration depends on the relevant masses and the elastic restoring and frictional forces. Whenever they exist, elastic forces such as Coulombic attraction inside molecules give the character of a resonance to the interaction, damped to a lesser or greater degree (radiation friction or Brehmstrahlung).

Figure 26 represents both real and imaginary components (K' and K") of a typical complex permittivity spectrum of a polar material containing space charges. We shall make a detailed analysis of this figure, starting from the high frequency side.

As we have seen, electrons of the inner atomic shells have critical frequencies of the order of 10^{19}Hz (X-ray range). Consequently, an e.m. field of frequency higher than 10^{19} (or of wavelength shorter than 1 Å) cannot excite any vibration in the atoms; hence, it has no polarizing effect on the material, which has, for this frequency, the same permittivity ε_o as a vacuum. (Point 1 of the figure).

Note that for $f > 10^{19}$ Hz, the relative permittivity (and also the refractive index since $K = n^2$) is smaller than unity. This means that the field propagates in the material with a velocity larger than that of light. This does not violate the law of relativity, since the velocity in question is a phase velocity.

If the frequency is lower than the resonant frequency of the inner electrons, these electrons can "feel" the electric component of the e.m. field, and they vibrate with the field. By so doing, they polarize the material, which raises the relative permittivity above unity.

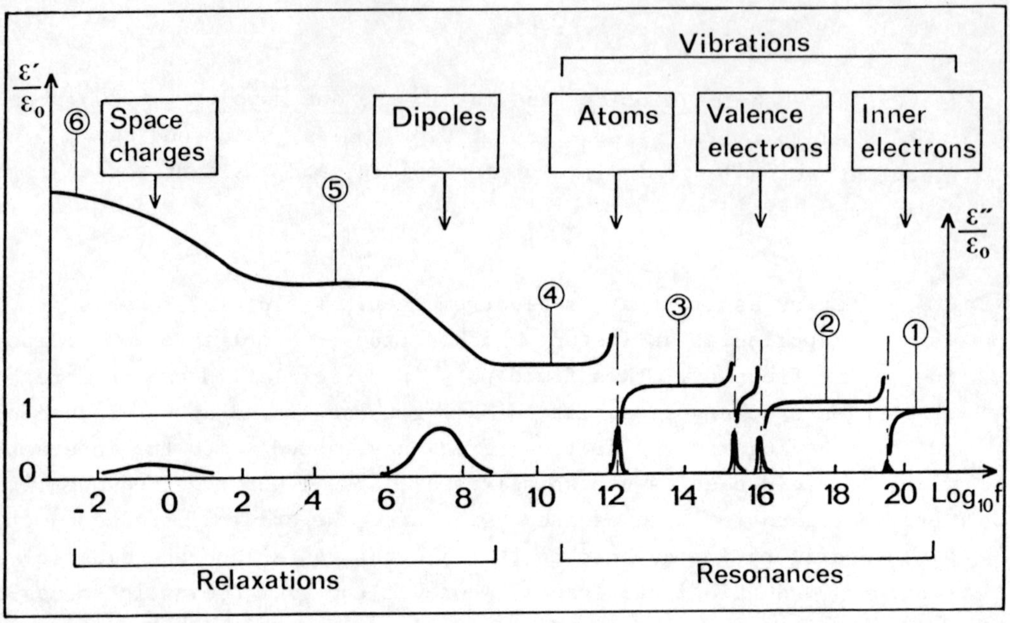

Fig.26 - The various types of interaction between the e.m. field and matter, and the relevant relative permittivities.

This is seen at point 2 of the figure.

Now, if the frequency of the e.m. field is lower than the resonant frequency of the valence electrons which is in the range 3×10^{14} to 3×10^{15} Hz (i.e. in the optical range from the u.v. (0.1 μm) to the near i.r. (1 μm)), these electrons also take part in the dielectric polarization, and their contribution again raises the permittivity. This is seen at point 3 on the figure.

The same type of "resonance" process occurs at the frequencies of atomic vibrations in molecules and crystals, in the range 10^{12} to 3×10^{13} Hz (i.e. in the far infrared spectral range between 10 and 300 μm). This is seen at point 4 of the figure.

In all the processes mentioned so far, the charges affected by the field can be considered to be attracted towards their central position by forces which are proportional to their displacements, i.e. by linear elastic forces. This mechanical approach of an electronic <u>resonance</u> is only approximate, since electrons cannot be treated properly by classical mechanics. Quantitative treatments of these processes require the formalism of quantum mechanics. However, the quantum numbers of these systems are usually so large that a classical resonance model including a friction term (to account for radiation damping) gives a fair description of these interactions (correspondence principle).

If the frequency of the applied field is lower than that of atomic vibrations, however, a new type of interaction may appear, in which the restoring forces are not elastic as in the case of direct interactions between charged particles, but viscous in character, as a result of an irreversible thermodynamic process. The sluggish collective orientation of dipoles or the accumulation of an ionic space charge near the electrodes when a field is applied or cut off is an example of such a process which is known as "relaxation".

Dipole relaxation, which is the time-dependent polarization due to the orientation of dipoles (cf. Chapter IV) is treated in Sections VII.2 to VII.5.

Interface polarization and relaxation in heterogeneous materials (Maxwell-Wagner effect), corresponds to the evolution from the "capacitive" voltage distribution of the short time (or high frequency) heterogeneous permittivity to the "resistive" distribution of the long time (or low frequency) heterogeneous resistivity. This is discussed in Section VII.7.

Space charge polarization and relaxation occurs in materials containing carriers which do not recombine at the electrodes, and therefore behave, in a low frequency a.c. field, as macroscopic dipoles which reverse their direction each half period. This will be discussed in Section VII.9.

VI.3. The response of a linear material to a variable field

First consider a rectangular pulse of field of amplitude E, applied to a linear material between times θ and $\theta + d\theta$ (Fig.27).

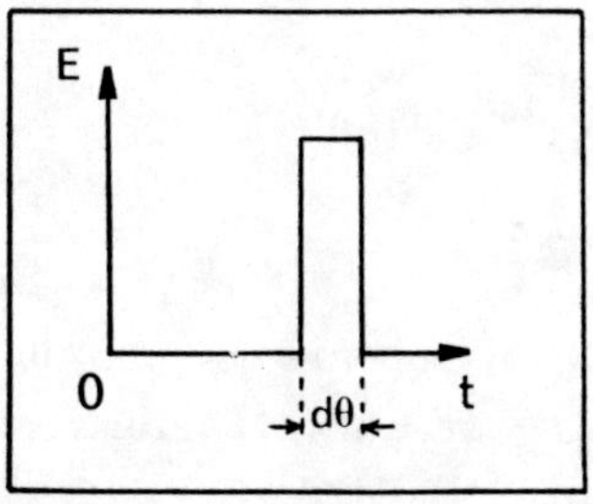

Fig.27 - A pulse of applied field.

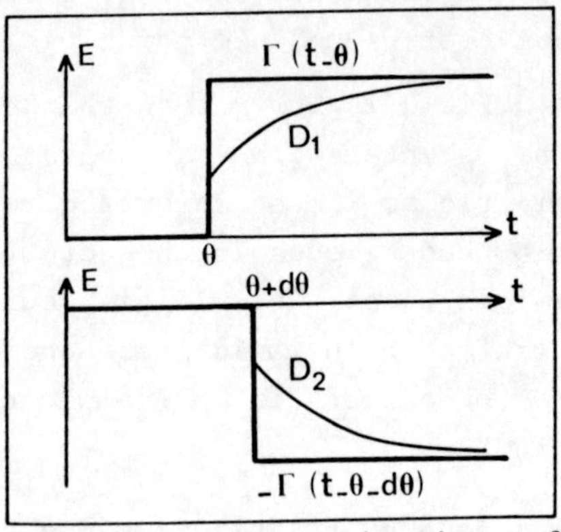

Fig.28- Decomposition of E(t) and D(t).

As can be seen in Figure 28, the pulse of applied field can be expressed as

$$E(t) = \Big[\Gamma(t-\theta) - \Gamma(t-\theta-d\theta)\Big]E \qquad \text{(VI.19)}$$

where $\Gamma(u)$ is the step function of unit amplitude.

From the superposition principle, the displacement D corresponding to the pulse is the sum of the displacement D_1 due to the leading edge and D_2 due to the trailing edge:

$$D_1(t) = \varepsilon_\infty E\,\Gamma(t-\theta) + (\varepsilon_s-\varepsilon_\infty)f(t-\theta)E \qquad \text{(VI.20)}$$

$$D_2(t) = -\varepsilon_\infty E\,\Gamma(t-\theta-d\theta)-(\varepsilon_s-\varepsilon_\infty)f(t-\theta-d\theta)E \qquad \text{(VI.21)}$$

In these equations, the first terms on the righthand sides refer to instantaneous polarization, of electronic origin, which appears immediately on application of the field, whereas the second terms refer to the sluggish polarization which is due to dipole orientation, or any other frictional process. The function $f(u)$ characterizes the sluggishness of the delayed polarization. It increases from $f(0) = 0$ to $f(\infty)=1$.

The system is linear and the displacement D due to the pulse $E(t)$ at t is:

$$D(t) = D_1+D_2 = \varepsilon_\infty E(t) + (\varepsilon_s-\varepsilon_\infty)f'\,E\,d\theta \qquad \text{(VI.22)}$$

with $E(t) = 0$ for $t \geqslant \theta + d\theta$.

The derivative $f'(u)$ of $f(u)$, which appears in the above equation, is a <u>decreasing</u> function, vanishing at infinite time, which is usually called the <u>decay function</u>, and denoted by Φ.

If the field is a continuous function of time, applied at $t = -\infty$, the displacement at instant t is:

$$D(t) = \varepsilon_\infty E(t) + (\varepsilon_s - \varepsilon_\infty) \int_{-\infty}^{t} \Phi(t-\theta) E(\theta) d\theta \tag{VI.23}$$

and if we use the new variable

$$u = t - \theta \qquad (d\theta = -du)$$

D(t) takes the form

$$D(t) = \varepsilon_\infty E(t) + (\varepsilon_s - \varepsilon_\infty) \int_{0}^{\infty} \Phi(u) E(t-u) du \tag{VI.24}$$

In this form, the integral is the convolution product of Φ and E, usually denoted by

$$\Phi(t) * E(t) \tag{VI.25}$$

VI.4. Case of an a.c. field. Kramers - Krönig relations

We assume that the applied field is sinusoidal, with amplitude E_0 and angular frequency $\omega = 2\pi f$:

$$E(t) = E_0 \cos \omega t \tag{VI.26}$$

Hence, using this formula in eqn.(VI.24), we have:

$$D(t) = \varepsilon_\infty E_0 \cos \omega t + (\varepsilon_s - \varepsilon_\infty) E_0 \int_{0}^{\infty} \Phi(u) \cos(\omega t - \omega u) du \tag{VI.27}$$

We deal here with times t which are much longer than the transient, so that D(t) finally becomes sinusoidal with the same angular frequency ω, but with a phase shift φ with respect to the applied field:

$$D(t \to \infty) = D \cos(\omega t - \varphi) = D_0 \cos \varphi \cos \omega t + D_0 \sin \varphi \sin \omega t \tag{VI.28}$$

From eqn.(VI.27), we obtain another expression for $D(t \to \infty)$:

$$D(t \to \infty) = \varepsilon_\infty E_0 \cos\omega t + (\varepsilon_s - \varepsilon_\infty) E_0 \int_{0}^{\infty} \Phi(u)(\cos\omega t \cos\omega u + \sin\omega t \sin\omega u) du \tag{VI.29}$$

or

$$D(t \to \infty) = \left[\varepsilon \ +(\varepsilon_s-\varepsilon_\infty) \int_0^\infty \Phi(u)\ \cos\omega u\ du\right] E_o \cos\omega t +$$
$$\left[(\varepsilon_s-\varepsilon_\infty) \int_0^\infty \Phi(u) \sin\omega t\right] E_o \ \sin\omega t \qquad \text{(VI.30)}$$

By solution of eqns (VI.28) and (VI.30), we obtain:

$$D_o \cos\varphi = \left[\varepsilon_\infty +(\varepsilon_s-\varepsilon_\infty) \int_0^\infty \Phi(u)\ \cos\omega u\ du\right] E_o \qquad \text{(VI.31)}$$

$$D_o \sin\varphi = \left[\ (\varepsilon_s-\varepsilon_\infty) \int_0^\infty \Phi(u)\ \sin\omega u\ du\right] E_o \qquad \text{(VI.32)}$$

But, by definition of the complex permittivity,

$$D = \varepsilon^* E = (\varepsilon' - i\varepsilon'')E \qquad \text{(VI.33)}$$

Hence, the "in phase" component of D corresponds to ε' and the "out of phase" component to ε''. It follows that

$$\varepsilon' = \varepsilon_\infty +(\varepsilon_s-\varepsilon_\infty) \int_0^\infty \Phi(u)\ \cos\omega u\ du \qquad \text{(VI.34)}$$

$$\varepsilon'' = (\varepsilon_s-\varepsilon_\infty) \int_0 \Phi(u)\ \sin\omega u\ du \qquad \text{(VI.35)}$$

The equations (VI.34) and (VI.35) show that both the real and imaginary parts of the permittivity depend on the same decay function Φ, which can be written as a Fourier transforms:

$$\Phi(u) = \frac{2}{\pi} \int_0^\infty \frac{\varepsilon'(x) - \varepsilon_\infty}{\varepsilon_s - \varepsilon_\infty} \cos ux\ dx \qquad \text{(VI.36)}$$

$$= \frac{2}{\pi} \int_0^\infty \frac{\varepsilon''(x)}{\varepsilon_s-\varepsilon_\infty} \sin ux\ dx \qquad \text{(VI.37)}$$

Consequently, the two spectra are interrelated, and the relation between them can be derived by substituting $\Phi(u)$, as given by eqn. (VI.37), in eqn.(VI.34):

$$\varepsilon'(\omega) - \varepsilon_\infty = \frac{2}{\pi} \int_0^\infty \left[\int_0^\infty \varepsilon''(x)\ \sin x u\ dx\right] \cos\omega u\ du \qquad \text{(VI.38)}$$

Changing the order of the integration gives

$$\varepsilon'(\omega) - \varepsilon_\infty = \frac{2}{\pi}\int_0^\infty \left[\int_0^\infty \sin x u \cos \omega u \, du\right] \varepsilon''(x) \, dx \qquad \text{(VI.39)}$$

The integral in the square brackets is

$$\int_0^\infty \sin xu \cos \omega u \, du = \frac{x}{x^2 - \omega^2} \qquad \text{(VI.40)}$$

so that

$$\boxed{\varepsilon'(\omega) - \varepsilon_\infty = \frac{2}{\pi}\int_0^\infty \frac{x\varepsilon''(x)}{x^2 - \omega^2} \, dx} \qquad \text{(VI.41)}$$

In a similar way, combination of eqns.(VI.35) and (VI.36) gives

$$\boxed{\varepsilon''(\omega) = \frac{2\omega}{\pi}\int_0^\infty \frac{\varepsilon'(x) - \varepsilon_\infty}{x^2 - \omega^2} \, dx} \qquad \text{(VI,42)}$$

Equations (VI.41) and (VI.42) constitute the Kramers-Krönig relations. They are valid for any type of dispersion, and permit one of the spectra to be obtained provided that the other has been measured throughout the complete spectral range. If we let $\omega = 0$ in eqn.(VI.41), we have:

$$\varepsilon_s - \varepsilon_\infty = \frac{2}{\pi}\int_0^\infty \varepsilon''(x) \frac{dx}{x}$$

or

$$\int_0^\infty \varepsilon''(\omega) \frac{d\omega}{\omega} = \frac{\pi}{2}(\varepsilon_s - \varepsilon_\infty) \qquad \text{(VI.43)}$$

This shows that the total area under the curve in a plot of ε'' vs. $\log \omega$ is simply related to the extreme values of the permittivity, irrespective of the dispersion mechanism.

VII. RELAXATIONS

VII.1. Introductory remarks

Relaxation processes are probably the most important of the interactions between fields and matter, at least from the standpoint of the physics of dielectrics.

We consider a collection of identical dipoles at temperature T in a constant, uniform applied field, as described in Chapter IV. According to the Langevin theory developed in Section IV.1, the interacting dipoles may assume any direction in space, but their steady state statistical orientation is such that

$$< \cos\theta > = \mathcal{L}(y) \sim \frac{1}{3}\, y,$$

where $\mathcal{L}(y)$ is the Langevin function of the variable $y = \mu E/kT$.

If the field is now abruptly cut off, all the torques due to the external field on the dipoles vanish instantly, and, by means of multiple collisions, the statistical orientation of the dipole system slowly disappears. The value of $< \cos\theta >$ decreases from $\mathcal{L}(y)$ to 0 with a characteristic time constant τ . This time constant is called the relaxation time constant. It is the same as that required by the initially isotropic system to become oriented after application of a step function of applied field.

In the theory he has developed in his famous book "Polar Molecules", P. Debye has extended the Langevin theory of dipole orientation in a constant field to the case of a varying field. In particular, if a constant field E applied for $t < 0$ is abruptly removed at $t = 0$, the theory describes the subsequent behaviour of the system for $t > 0$. It shows that the constant Boltzmann factor $\exp(y \cos\theta)$ of the Langevin theory becomes a time-dependent weighting factor:

$$f(t) = \exp\left[y \cos\theta\, \varphi(t) \right] \qquad \text{(VII.1)}$$

Strickly speaking, $\varphi(t)$ is a decaying function involving E , but for the usual case $y \ll 1$, it reduces to:

$$\varphi_0(t) = \exp(-t/\tau) \qquad \text{(V11.2)}$$

where the relaxation time τ is related to the internal friction coefficient ζ by the equation

$$\tau = \frac{\zeta}{2kT} \qquad \text{(VII.3)}$$

In the case of spherical or nearly spherical molecules, Stokes law can be used to relate the internal friction coefficient ζ to the molecular radius a and the viscosity η :

$$\zeta = 8\pi \eta a^3 \qquad \text{(VII.4)}$$

and since the viscosity η is a thermally activated quantity,

$$\eta = \eta_o \exp(\frac{U}{kT}), \qquad \text{(VII.5)}$$

τ takes the form

$$\tau(T) = \frac{8\pi\eta_o a^3}{2kT} \exp(\frac{U}{kT}) \qquad \text{(VII.6)}$$

Roughly speaking, the relaxation times for dipole orientation at room temperature are between 10^{-10} s, for small dipoles diluted in a solvent of low viscosity and more than 10^{-4} s for big dipoles in a viscous medium. There are also dipole relaxations in crystals (namely, the relaxations associated with pairs of lattice vacancies) and in polymers (such as the α, β and γ bands in polyethylene).

Returning to $f(t)$ in eqn.(VII.1), the dynamical variation of $<\cos\theta>$ is given by:

$$<\cos\theta(t)>_{sp.aver} = \frac{\int \cos\theta \exp\left[y \cos\theta \varphi(t)\right] d\Omega}{\int \exp\left[y \cos\theta\, \varphi(t)\right] d\Omega} \qquad \text{(VII.7)}$$

This is a complicated function of time. However, if $y \ll 1$, the exponentials can be replaced by their first order expansions, with $\varphi(t) = \varphi_o(t)$:

$$\exp\left[y \cos\theta\, \varphi(t)\right] \simeq 1 + y \cos\theta\, \varphi_o(t)$$

and in this case,

$$<\cos\theta(t)>_{sp.aver} \simeq \frac{1}{3} y\, \varphi_o(t) = \frac{1}{3} y \exp(- t/\tau) \qquad \text{(VII.8)}$$

Problem

The case when y is not negligible.

If y is not very small with respect to unity, the Debye theory predicts that $\varphi(t)$ is not the simple exponential $\varphi_o(t)$ but is a more complicated function which can be defined as the solution of the differential equation:

$$\tau \frac{d\varphi}{dt} + \left[1 - \frac{y}{2} <\sin^2\theta(t)>\right]\varphi = 0$$

where $\langle \sin^2\theta(t) \rangle$ is time-dependent.

1. Write $\langle \sin^2\theta(t) \rangle$, assuming that it varies exponentially between its initial and its final values.
2. Using for $\langle \sin^2\theta(t) \rangle$ the time variation which is found above, write an approximation for $\langle \cos\theta \rangle$ based on eqn.(VII.7), and compare the result with eqn.(VII.8).

VII.2. Mechanical analogue of a relaxation

We return to the analogy between these sluggish processes and the motion of a charge controlled by friction or viscous drag, described previously.

The dynamical equation for a charge in a field E and undergoing a frictional force fv proportional to its velocity v is

$$m \frac{dv}{dt} = eE - fv \tag{VII.9}$$

or

$$\frac{dv}{dt} = \frac{1}{\tau}(v_s - v) \tag{VII.10}$$

where

$$\tau = \frac{m}{f} \quad \text{and} \quad v_s = \frac{eE}{f} \quad \text{is the final velocity.}$$

Equation (VII.10) can readily be integrated to give

$$v(t) = v_s \left[1 - \exp(-t/\tau) \right] \tag{VII.11}$$

The orientational polarization of a system of dipoles submitted to a step function of field obeys exactly the same type of equation, which means that:

$$P_{or}(t) = (P_s - P_\infty) \left[1 - \exp(-t/\tau) \right] \tag{VII.12}$$

so that, for the total displacement:

$$D(t) = D_\infty + (D_s - D_\infty) \left[1 - \exp(-t/\tau) \right] \tag{VII.13}$$

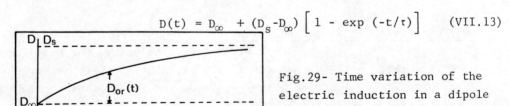

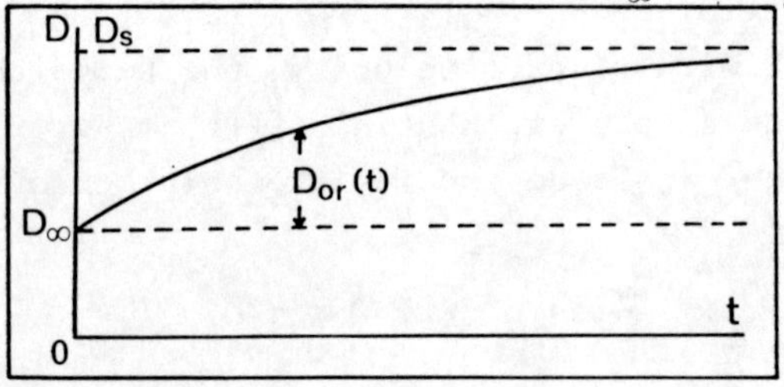

Fig.29- Time variation of the electric induction in a dipole system.

Figure 29 shows the time variation of D .

VII.3. Advanced formalism. Definitions and theorems

First, we shall give general, rigorous definitions and theorems without proof, concerning the frequency response of a linear system. Proofs can be found in textbooks on applied mathematics dealing with Laplace transforms, and also in some of the general references given in this book, e.g. in "Dielectric Relaxation" by V. Daniel.

We shall then apply the theorems to the case of simple dipole relaxation, mainly to illustrate their elegance and power. However, we shall abandon this rather abstract formalism later, and replace it by a less rigorous but more elementary approach, by means of which more complex systems, such as interfacial and space-charge relaxations, resonances, etc. can be treated directly, in an approximate, though quite acceptable manner.

Definition

The frequency response $F(\omega)$ of any characteristic physical quantity (polarization, displacement, etc.) in a linear system is the magnitude of the relevant response of the system at angular frequency ω, when it is submitted to an excitation (for instance an applied field) at the same frequency.

Theorem I

$F(\omega)$ is the imaginary Laplace transform of the "Decay function" for the particular quantity.

$$F(\omega) = \int_0^\infty \Phi(t) e^{-i\omega t}\, dt \qquad \text{(VII.14)}$$

Theorem II

The Laplace transform of a convolution product is a simple product.

VII.4. Application to dipole relaxation - Debye relation

It has previously been shown that the permittivity due to the dipole orientation in a variable field $E(t)$ is such that

$$D_{or}(t) = \varepsilon_{or}(t) * E(t) \qquad \text{(VII.14)}$$

According to Theorem II, this implies

$$D(\omega) = \varepsilon^*_{or}(\omega)\, E(\omega) \tag{VII.15}$$

We recall that dipole orientation has been compared to a first order mechanical system (i.e. a system in which polarization and displacement vary exponentially with time after the application (or the cut-off) of a stepfunction of field). Consequently, the characteristic decay function Φ is an exponential of the form $\exp(-t/\tau)$. In order to keep the Laplace transform dimensionless, however, we write the decay function in the form

$$\Phi(t) = \frac{1}{\tau} \exp(-t/\tau) \tag{VII.16}$$

Thus, from Theorem I,

$$\varepsilon^*_{or}(\omega) = \frac{\varepsilon_s - \varepsilon_\infty}{\tau} \int_0^\infty e^{-t/\tau - i\omega t}\, dt = \frac{\varepsilon_s - \varepsilon_\infty}{1 + i\omega\tau}$$

and

$$\boxed{\varepsilon^*(\omega) = \varepsilon_\infty + \frac{\varepsilon_s - \varepsilon_\infty}{1 + i\omega\tau}} \tag{VII.17}$$

This is the famous Debye relation in its simplest form, where the conductivity of the material is neglected.

It is interesting to note that the frequency response of a relaxing system, such as the first-order mechanical system of eqn. (VII.10), can be obtained by simply replacing the variable (here v) by the oscillating quantity $v_o e^{i\omega t}$. From this, it follows that $\frac{dv}{dt} = i\omega v$, so that

$$v = \frac{v_s}{1 + i\omega\tau} \tag{VII.18}$$

Application of this simple technique to the displacement in eqn. (VII.13) gives the Debye relation directly.

We can calculate the real and imaginary parts ε' and ε'' of $\varepsilon^* = \varepsilon' - i\,\varepsilon''$:

$$\varepsilon'(\omega) = \varepsilon_\infty + \frac{\varepsilon_s - \varepsilon_\infty}{1 + \omega^2\tau^2} \tag{VII.19}$$

$$\varepsilon''(\omega) = (\varepsilon_s - \varepsilon_\infty)\, \frac{\omega\tau}{1 + \omega^2\tau^2} \tag{VII.20}$$

From these equations, the curves of Fig.30 are obtained

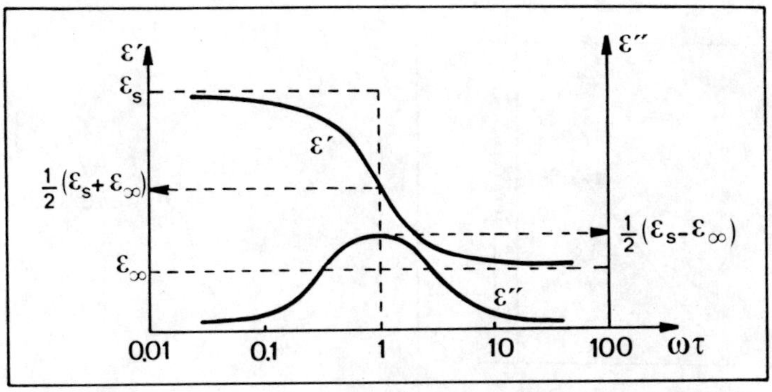

Fig.30- ε' and ε'' vs. $\log(\omega\tau)$

Problem

Show that the width at half-height of the loss peak $\varepsilon''(\omega)$ given by eqn.(VII.20) is 1.144 decades.

Problem

Using Maxwell's eqns.(VI.16) and (VI.17), find the absorption coefficient α of a dipole system which obeys the Debye model. Discuss the variation of α with ω, and, in particular, the limiting cases $\omega \to 0$ and $\omega \to \infty$. Discuss the paradox raised by the limit $\omega \to \infty$.

VII.5. The $\varepsilon''(\varepsilon')$ representation (Argand diagram)

Another graphical representation which is of considerable practical importance involves plotting ε' (or K') versus ε''. The function $\varepsilon''(\varepsilon')$ can, of course, be obtained by elimination of ω between the equations for $\varepsilon'(\omega)$ and $\varepsilon''(\omega)$, and, in the case of the simple dipole relaxation, the functional relation obtained in this way can be shown to be a circle.

The same result can be obtained without calculation, simply by looking at the complex quantity ε^* as represented by a vector in the plane $(\varepsilon',\varepsilon'')$. This vector is the resultant of the real quantity ε_∞ and the complex quantity $(\varepsilon_s-\varepsilon_\infty).(1+i\omega\tau)^{-1}$. The complex denominator $(1+i\omega\tau)$ of this component is represented by a vector connecting the origin to a point on the line $\varepsilon' = 1$. Its inverse $(1+i\omega\tau)^{-1}$ is therefore represented by a vector connecting the origin to a point on a semi-circle centered on the real axis, as shown in Figure 31.

ε^* is now represented by the semi-circle of radius $(\varepsilon_s-\varepsilon_\infty)/2$ centred at $\varepsilon' = (\varepsilon_\infty+\varepsilon_s)/2$. The top of this semi-circle corresponds to $\omega\tau = 1$.

*

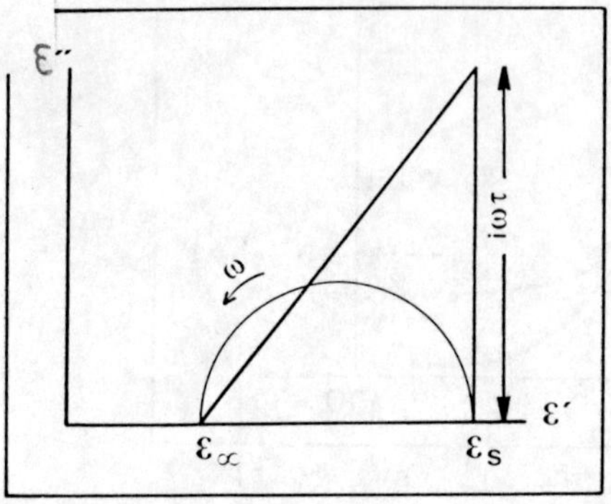

Fig.31- Construction of the Cole-Cole diagram.

VII.6. Corrections to the Debye theory

The treatment which was given above implies a number of simplifying assumptions, namely:

1. The local field is not different from the applied field
2. The conductivity of the material is negligible
3. All the dipoles have the same relaxation time τ .

We now discuss the corrections which have to be made to the Debye relation whenever the above assumptions are not fully justified.

1. Influence of the local field

As we have seen, assumption 1 is valid only in the case of dilute gases. In condensed matter, the Lorentz field correction gives:

$$E_{loc} = E + \frac{P}{3\varepsilon_o} = \frac{\varepsilon + 2\varepsilon_o}{3\varepsilon_o} E \ , \qquad \text{(VII.21)}$$

so that the displacement $D = \varepsilon_o E + P$, expressed as a function of the local field becomes:

$$D = \varepsilon_o \frac{3\varepsilon}{\varepsilon + 2\varepsilon_o} E_{loc} \qquad \text{(VII.22)}$$

Within the frame of the Lorentz local field model, eqn.(VII.22) is equally valid for frequencies at which the permittivity is complex as for the very high and very low frequencies corresponding to the extreme real values ε_∞ and ε_s, respectively. Consequently, to account for the local field, we have to introduce the above value of D into the relaxation equation:

$$D^*(\omega) = D_\infty + \frac{D_s - D_\infty}{1 + i\omega\tau} \qquad (VII.23)$$

This gives :

$$\frac{\varepsilon^*}{\varepsilon^* + 2\varepsilon_0} = \frac{\varepsilon_\infty}{\varepsilon_\infty + 2\varepsilon_0} - \left(\frac{\varepsilon_s}{\varepsilon_s + 2\varepsilon_0} - \frac{\varepsilon_\infty}{\varepsilon_\infty + 2\varepsilon_0}\right)\frac{1}{1 + i\omega\tau} \qquad (VII.24)$$

or

$$\frac{\varepsilon_\infty + 2\varepsilon_0}{\varepsilon^* + 2\varepsilon_0} = 1 - \frac{\varepsilon_s - \varepsilon_\infty}{(\varepsilon_s + 2\varepsilon_0)(1 + i\omega\tau)}$$

From this, ε^* can be written in the form of the Debye relation :

$$\varepsilon^*(\omega) = \varepsilon_\infty + \frac{\varepsilon_s - \varepsilon_\infty}{1 + i\omega\tau'} \qquad (VII.25)$$

with

$$\tau' = \frac{\varepsilon_s + 2\varepsilon_0}{\varepsilon_\infty + 2\varepsilon_0} \qquad (VII.26)$$

The measured relaxation time, which is the reciprocal of the angle at the top of the Cole-Cole diagram, is therefore higher than the real relaxation time. For example, in the case of nitrobenzene, $\varepsilon_s = 34\varepsilon_0$, $\varepsilon_\infty = 2.5\varepsilon_0$ so that $\tau'/\tau = \frac{36}{4.5} = 8$. This, however is illustrative rather than rigorous, since the Lorentz field is not an acceptable local field for condensed polar materials.

In some anisotropic materials such as liquid crystals, the measured relaxation time τ' can be up to three orders of magnitude longer than the relaxation time measured in a dilute, non-polar solvent. This sort of problem is now under investigation.

Problem 1

Use the Onsager model to find a better expression for the local field at low frequencies, and discuss the relevant correction to the Debye formula.

2. Influence of the d.c. conductivity

It was seen in Section VI.1 that the contribution to the permittivity ε^* of the conductivity σ due to free charges is $(-i\sigma/\omega)$.

Since a conducting material can be regarded as a non-conducting dielectric with a resistance in parallel, the equation describing the complex permittivity of a conducting, polar material is:

$$\varepsilon^*(\omega) = \varepsilon_\infty + \frac{\varepsilon_s - \varepsilon_\infty}{1 + i\omega\tau} - i\,\frac{\sigma}{\omega} \qquad \text{(VII.27)}$$

The influence of the last term of eqn.(VII.27) on the Cole-Cole diagram can be seen in Fig. 32 . Of course the larger the conductivity, the further the actual diagram departs from the Cole-Cole semi-circle.

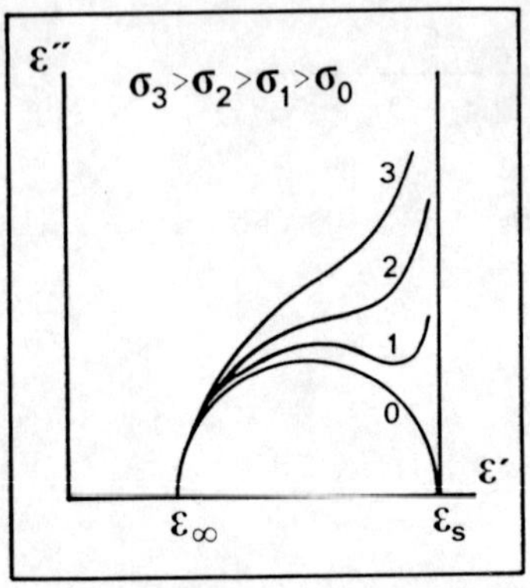

Fig. 32- Influence of the d.c. conductivity σ on the Argand diagram of a dipole system.

The alternative graphical representation of plotting the logarithm of ε'' against the log. of the frequency (Fig. 33) permits the a.c. conductivity associated with the relaxing dipoles to be distinguished easily from the d.c. conductivity σ due to the free charges. The imaginary part of ε^* given by eqn.(VII.27) is:

$$\varepsilon'' = (\varepsilon_s - \varepsilon_\infty)\,\frac{\omega\tau}{1 + \omega^2\tau^2} + \frac{\sigma}{\omega} \qquad \text{(VII.28)}$$

and from this expression, we can see that:

for $\omega\tau \ll 1$, $\varepsilon'' \simeq \frac{\sigma}{\omega}$ (range I of Fig. 33)

for $\omega\tau \gg 1$, $\varepsilon'' \simeq \frac{\varepsilon_s - \varepsilon_\infty}{\omega\tau}$ (range II of Fig. 33)

In the intermediate range $(\omega\tau \simeq 1)$,

$$\varepsilon'' \simeq \frac{\varepsilon_s - \varepsilon_\infty}{2}\,\omega\tau + \sigma\tau \qquad \text{(VII.30)}$$

This result is shown in Fig. 33.

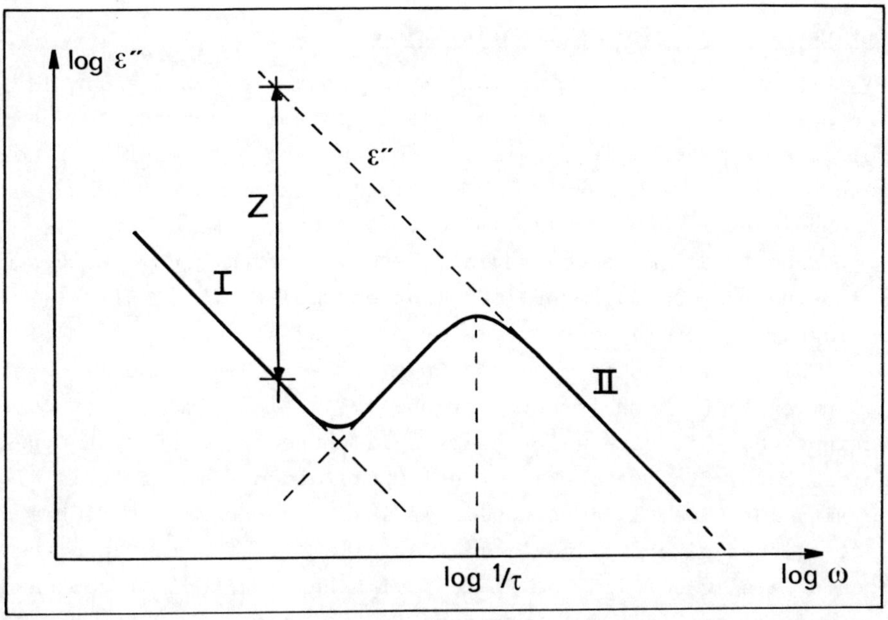

Fig.33- Plot of log ε'' vs. log ω , according to eqn.(VII.27)

I zone $\omega\tau \ll 1$

$$\varepsilon''_{I} = \frac{\sigma}{\omega} \Rightarrow \log \varepsilon''_{I} = \log \sigma - \log \omega$$

II zone $\omega\tau \gg 1$

$$\varepsilon''_{II} = \frac{\varepsilon_s - \varepsilon_\infty}{\omega\tau} \Rightarrow \log \varepsilon''_{II} = \log \frac{\varepsilon_s - \varepsilon_\infty}{\tau} - \log \omega$$

Difference : $Z = \log \varepsilon''_{II} - \log \varepsilon''_{I} = \log \frac{\varepsilon_s - \varepsilon_\infty}{\sigma\tau}$

The second term ($\sigma\tau$) of eqn.(VII.30) is usually small with respect to the first. For instance, with the typical values, $\sigma = 10^{-8}$ s and $\tau = 10^{-10}$ s, $\sigma\tau = 10^{-18}$. This is quite small compared to first term, which is of the order of 10^{-11}, and can be neglected. In Fig.33, where the coordinates are $\log \varepsilon''$ and $\log \omega$, the vertical shift Z between the two branches with slope (-1) is, therefore, $\log \frac{\varepsilon_s - \varepsilon_\infty}{\sigma\tau}$.

The relaxation time τ is measured (and eventually corrected) as the reciprocal of the angular frequency at the maximum of ε'', and $(\varepsilon_s - \varepsilon_\infty)$ is known from the intercepts of the Cole-Cole diagram with the ε' axis. The conductivity σ can be deduced simply from the measured shift Z, which should be independent of ω.

3. Influence of multiple relaxation times

The $\varepsilon''(\varepsilon')$ diagram of many polar molecules in the liquid phase is actually a semi-circle of the Cole-Cole type, as predicted by the simple theory given above. This is, for instance, true of most dehydrated alcohols, and of solutions of symmetrical polar molecules, such as chlorobenzene, in a non-polar solvent (benzene, alkanes, etc.).

Many $\varepsilon''(\varepsilon')$ spectra deviate, however, from a Cole-Cole semi-circle. The maximum value of ε'' is then smaller than $\frac{1}{2}(\varepsilon_s - \varepsilon_\infty)$, and the diagram may or may not have a symmetry axis.

This deviation from a Cole-Cole semi-circle is usually explained by assuming that there is not just one relaxation time, but a continuous distribution. This occurs, of course, if there are different types of dipoles, each with its characteristic relaxation time. This is also the case with identical long molecules in which the permanent dipole moment is not aligned with the long molecular axis. If, for instance, the molecule is aligned with the field, only the longitudinal component of the dipole moment (Fig. 34) is active in the relaxation, and the molecule tends to rotate about a short molecular axis (a), with a long relaxation time, due to inertial and viscous forces. Conversely, if the molecule is perpendicular to the field, the transverse component $\mu_\perp$ of the dipole is active, so that the molecule relaxes by rotating rather quickly about

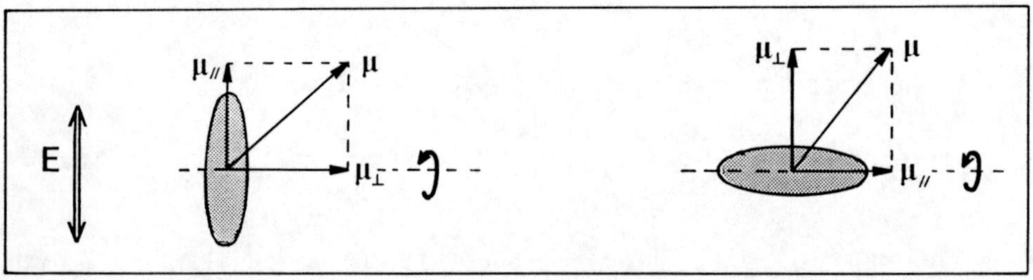

Fig.34 - a) slow relaxation about a transverse axis. b) fast relaxation about a longitudinal axis.

its long axis, since inertial and viscous forces are smaller in this configuration.

Since the molecules are randomly oriented with respect to the field, the relaxation times in a large collection of molecules are distributed between those of the two extreme cases which have just been considered. If we call $f(\tau)\,d\tau$ the relative number of molecules having a relaxation time between τ and $(\tau + d\tau)$, $f(\tau)$ is the normalized distribution function of the relaxation time.

For a given distribution function, there is a corresponding a complex permittivity

$$\varepsilon^*(\omega) = \varepsilon_\infty + (\varepsilon_s - \varepsilon_\infty) \int_0^\infty \frac{f(\tau)d\tau}{1 + i\omega\tau} \tag{VII.31}$$

The limits of the integral have no physical significance. They could be τ_{min} and τ_{max}, but using 0 and ∞ makes mathematical treatments simpler.

Conversely, given an Argand diagram, it is possible, in principle, to find the corresponding distribution function $f(\tau)$ by a Taylor-integral technique, and from the distribution function $f(\tau)$, it is possible to derive the ratios between the principal molecular axes.

However, the complexity of this calculation is not justified by the physical significance of the results, which neglect both the multipolar interactions between a molecule and its neighbours, and the possible interactions within a given molecule.

Elementary, semi-empirical methods can conveniently be used to deduce Argand spectra from given simple distributions such as:

a) bounded distributions of the form

$$f(\tau) = \begin{cases} A,\ B/\tau,\ \text{etc.} & \text{for} \quad \tau_1 \ll \tau \ll \tau_2 \\ 0 \ \text{outside} & \end{cases} \tag{VII.32}$$

b) normal (Gaussian) distributions in a log scale:

$$f(\tau) = C \exp\left[- (\ln \frac{\tau}{\tau_o})^2 \right] \tag{VII.33}$$

An acceptable fit can often be obtained by correctly choosing the parameters τ_1, τ_2, A, B, C, τ_o, etc.

Another test which can be made on a Argand diagram with its low and high frequency branch perpendicular to the ε' axis is to try to describe the relaxation in terms of a finite number of Debye relaxation processes. To make this clear, we consider the simplest example of two Debye relaxations:

$$\varepsilon^*(\omega) = \varepsilon_\infty + \frac{\varepsilon_1}{1 + i\omega\tau_1} + \frac{\varepsilon_2}{1 + i\omega\tau_2} \tag{VII.34}$$

where ε_1 and ε_2 obey the condition for vanishing frequency:

$$\varepsilon_s = \varepsilon_\infty + \varepsilon_1 + \varepsilon_2 \tag{VII.35}$$

The complete diagram corresponding to eqn.(VII.34) is constructed point by point in Fig.35, using:

$$\tan\alpha_1 = \omega\tau_1 \quad \text{and} \quad \tan\alpha_2 = \omega\tau_2$$

which implies that

$$\frac{\tan\alpha_2}{\tan\alpha_1} = \frac{\tau_2}{\tau_1} \qquad \forall\,\omega \tag{VII.36}$$

The diagram differs significantly from a semi-circle only if the two relaxation times are very different from one another.

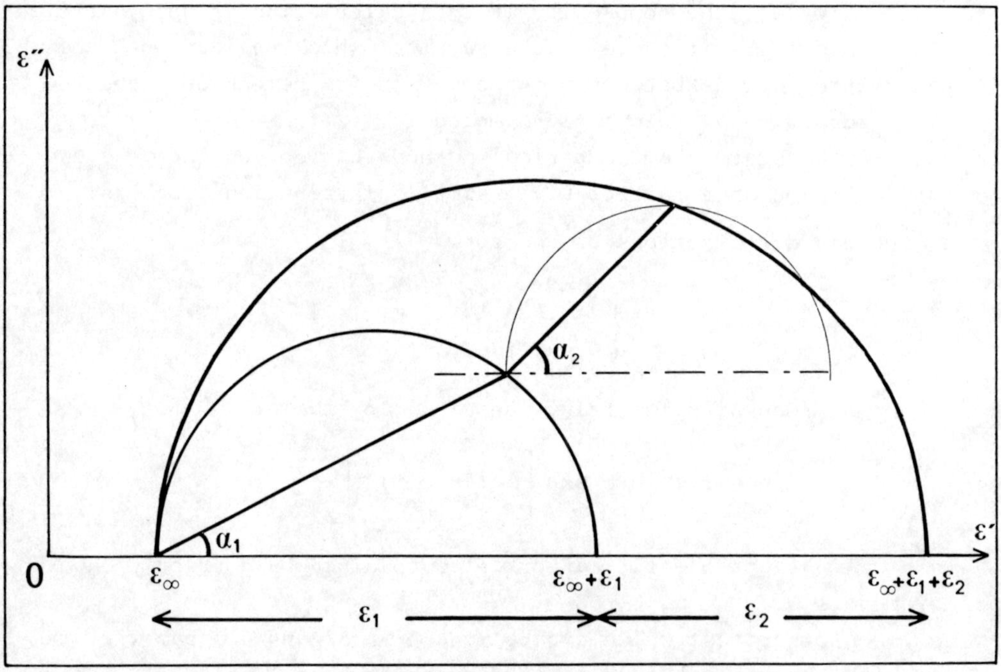

Fig.35- Construction of $\varepsilon''(\varepsilon')$ according to eqn: (VII.34).$(\tau_2/\tau_1=2)$

Conversely, assuming that ε_1 and τ_1 are known, equations (VII.34) and (VII.35) permit the evaluation of ε_2 and τ_2, assuming that the process under study is a double relaxation. Computer programs have been developed to decompose any Argand diagram in a small number of relaxation components.

Special diagrams. Cole and Cole arc

From the observation that the Argand diagram is often a circular arc centred below the ε' axis, Cole and Cole have proposed a modified Debye formula:

$$\varepsilon^*(\omega) = \varepsilon_\infty + \frac{\varepsilon_s - \varepsilon_\infty}{1 + (i\omega\tau)^{1-h}} \qquad \text{(VII.37)}$$

where h is a parameter characterizing the flattening of the diagram. If $h = 0$, eqn.(VII.37) reduces to the Debye relation. For $0 \leqslant h \leqslant 1$ it can easily be seen (Fig.36), by means of the inversion used before, that the Argand diagram is a circular arc centred below the ε' axis, at a distance $\frac{\varepsilon_s - \varepsilon_\infty}{2} \tan \frac{h\pi}{2}$ from this axis.

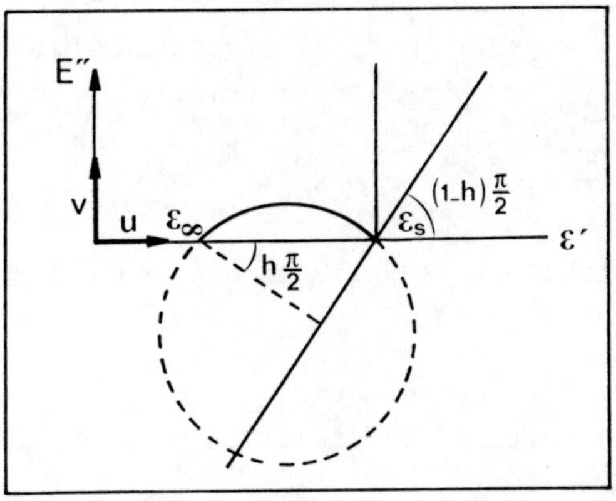

Fig.36 - Construction of the Cole-Cole arc diagram for eqn.(VII.37).

Problem 2

We suggest that the reader verifies the position of the circle which is given above, draws the arc for the particular value $h = 0.5$, and finds the particular distribution function $f(\tau)$ for this value of h.

Problem 3

Transient currents.

From section (VI.1), it results that any sample of a linear material in an a.c. field can be ascribed a complex admittance

$$y^* = i\omega C^*$$

where C^* is its complex capacitance $C^* = K^* C_o$, K^* the complex permittivity of the material, and C_o the "geometric" capacitance.

Using Heaviside's notations ($i\omega = p$), y^* becomes

$$y^*(p) = pC_o K^*(p)$$

Now, it results from Section VII.3 that the transient response to an applied voltage $V(t)$ is the Laplace transform of the simple product $y(p)U(p)$, where $U(p)$ is the "image" of $V(t)$.

For a step function of amplitude V, $U(p)$ reduces to $U(p)=V/p$, so that

$$i(t) \supset C_o V K^*(p)$$

means that the transient current for a step function of amplitude V is $C_o V$ times the Laplace transform of $K^*(p)$.

1. Transient in a dipolar material presenting Debye's relaxation.

If K^* is given by the Debye relation

$$K^*(p) = K_\infty + \frac{K_s - K_\infty}{1+p\tau},$$

show that the transient is a simple exponential decay with the time constant τ.

2. Transient for a non Debye material.

Now, assume that $K^*(p)$ is given by eqn.(VII.37):

$$K^*(p) = K_\infty + \frac{K_s - K_\infty}{1 + (p\overline{\tau})^{1-h}}$$

The Laplace transform of $F(p) = \left[1 + p^n\right]^{-1}$ is only known for $n = 1$ (case treated above) and $n = 1/2$. Derive the expression for $i(t)$ in this latter case, knowing that

$$\frac{1}{1+\sqrt{p}} \subset \frac{1}{\sqrt{\pi t}} - e^t \operatorname{erfc}(\sqrt{t})$$

with $\operatorname{erfc}(x) = 1 - \operatorname{erf}(x) = 1 - \frac{2}{\sqrt{\pi}} \int_0^x e^{-u^2} du$.

For any other value of h, $F(p)$ is not know in compact form, but it can be expanded in terms of $\frac{1}{p\overline{\tau}}$ if $p\overline{\tau} > 1$ $(t < \overline{\tau})$, or in terms of $p\overline{\tau}$ if $p\overline{\tau} < 1$ $(t > \overline{\tau})$.

3. Remembering that

$$p^{-m} \subset \frac{t^{m-1}}{\Gamma(m)},$$

calculate the expansions of $i(t)$ in terms of $t/\overline{\tau}$ for $t < \overline{\tau}$, and in terms of $\bar{\tau}/t$ for $t > \bar{\tau}$. To deduce the second expansion from the first, use the fact that changing h into (2-h) is the same as changing $p\bar{\tau}$ into $1/p\bar{\tau}$.

4. Show that the leading terms of the above expansions are proportional to $(t/\bar{\tau})^{-h}$ for $t \ll \bar{\tau}$, and to $(t/\bar{\tau})^{-2+h}$ for $t \gg \bar{\tau}$, except for $h = 0$ and $h = 1$ where the expansions are not valid.
Use this to represent - for some values of h $(0 \leqslant h \leqslant 1)$ - the reduced current

$$j = \frac{\bar{\tau}\, i}{C_0 V (K_s - K_\infty)}$$

versus the reduced time $\theta = t/\bar{\tau}$ on a log.log scale, in the interval $10^{-2} \leqslant \theta \leqslant 10^{2}$.

Problem 4

The Hamon approximation.

Hamon has proposed a simple relation to deduce the $K''(\omega)$ dispersion spectrum from the transient response $i(t)$ to a voltage step V (Proc. I.E.E.II 99 314 (1952)):

$$K''(\omega) = \frac{i(2/\pi\omega)}{\omega C_0 V}$$

where $i(2/\pi\omega)$ is the value of the transient current at time $t = 2/\pi\omega$.

1. Same as question 1 of the previous problem.

2. Plot on the same semilog. graph $K''/(K_s - K_\infty)$ versus $\log_{10}\omega\tau$ (for $10^{-3} \leqslant \omega\tau \leqslant 10^{3}$),

   a - using Debye's relation
   b - using eqn.(VII.37) with $h = 1/2$
   c - using Hamon's approximation with $i(t)$ of question 1.

3. Consider the function

$$f(\omega) = \frac{i(a/\omega)}{\omega C_0 V}$$

Using the fact that $d(\ell n\omega) = d\omega/\omega$, calculate the surface area under the curve $f(\log\omega)$.
Establish the validity of Hamon's approximation by showing that the value of a for which the area agrees with eqn.(VI.43) is $a = 2/\pi$.

Davidson and Cole skewed diagram

Quite often, the $\varepsilon''(\varepsilon')$ diagram is not symmetrical, and in this case, it may be fairly well represented by the simple analytic relation, proposed by Davidson and Cole :

$$\varepsilon^*(\omega) = \varepsilon_\infty + \frac{\varepsilon_s - \varepsilon_\infty}{(1 + i\omega\tau)^\alpha} \qquad \text{(VII.38)}$$

with

$$0 \leqslant \alpha \leqslant 1$$

For $\alpha = 1$, this reduces again to the Debye relation, but if α is smaller than unity, the diagram appears as the Figure 37.

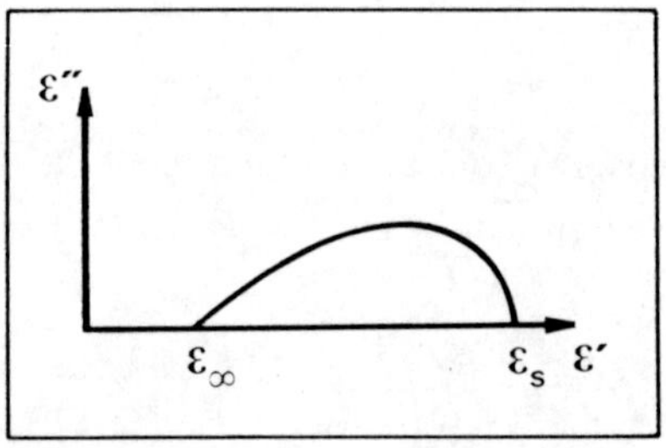

Fig.37- Davidson and Cole skewed diagram.

As an illustration, Figure 38 shows diagrams measured at various temperatures with Pyralene, which is a French trade-name for fluid mixtures of chlorinated diphenyls in chlorobenzene, similar to the arochlors. Although their industrial use is now strictly controlled since their thermal degradation may produce polluting agents, these fluids have been used extensively in the past, thanks to their high permittivity and insulating power, as impregnants for condensers and cables. The data of Fig.38, obtained in the author's laboratory, illustrate the skewed shape of the diagrams and the change of the relaxation times with temperature (cf. eqn. VII.6).

This illustrates the complementary relationship between frequency and temperature.

In fact, measuring ε^* at a suitable fixed frequency and different temperatures is equivalent to measuring ε^* at a fixed temperature and different frequencies, and both techniques should give the same kind of information, provided that a) conductivity due to free carriers remains negligible and b) no transition occurs in the sample over the temperature range under investigation.

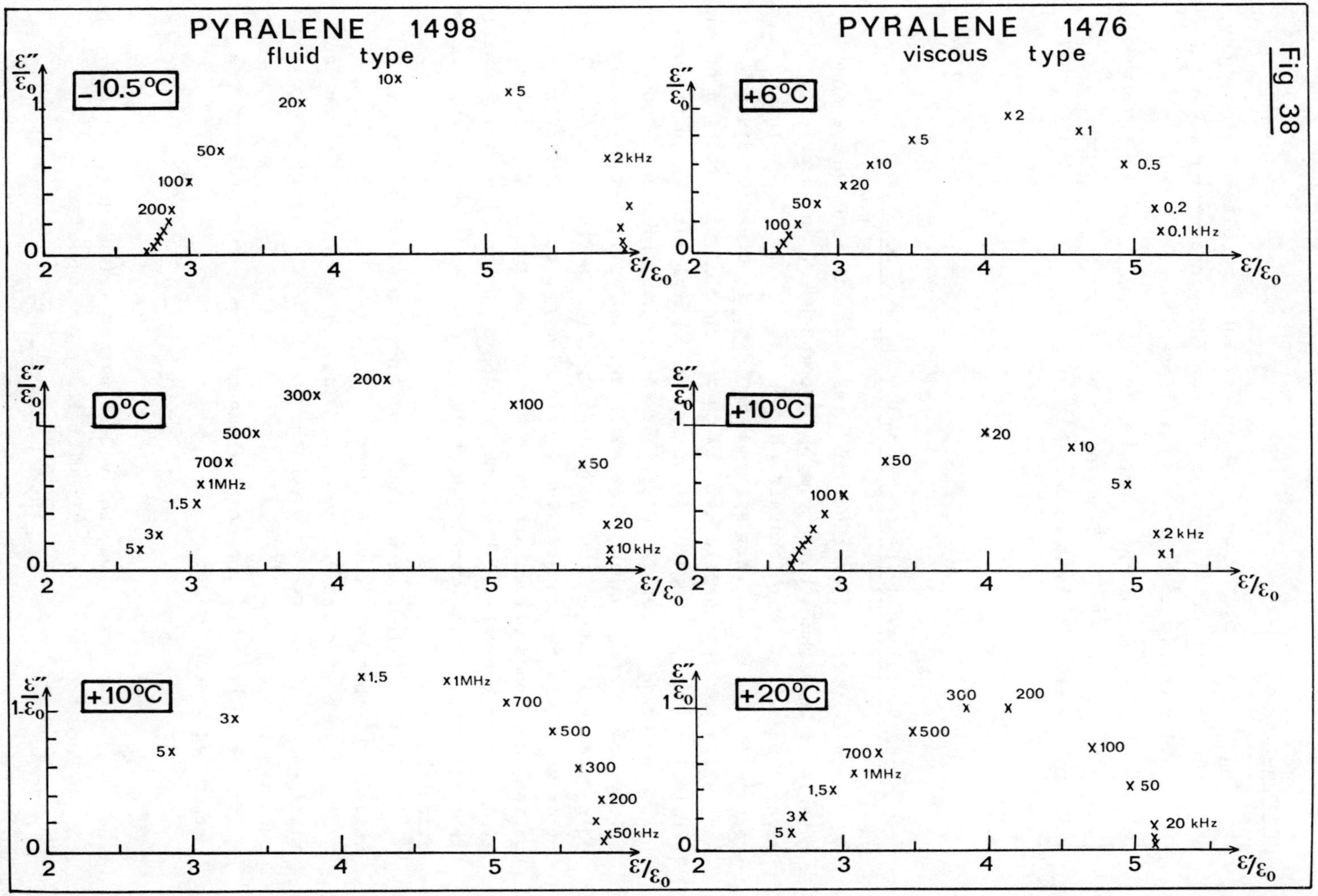
PYRALENE 1498
fluid type
PYRALENE 1476
viscous type
-10.5°C
0°C
+10°C
+6°C
+10°C
+20°C
ε''/ε0
ε'/ε0
2kHz
10kHz
50kHz
0.1kHz
2kHz
20kHz
1MHz

Fig 38

Problem 5

1. Find the best fit between the experimental results of the Figure 38 and the relation of Davidson and Cole. The proper value of α can be obtained from simple geometric considerations.

2. Draw a diagram for the special value $\alpha = 0.5$. On this diagram, mark the point where $\omega\tau = 1$. Find the value of $\omega\tau$ at the top of the diagram, where ε'' is maximum.

VII.7 Interfacial relaxation - Maxwell-Wagner effect

This is the relaxation which takes places in heterogeneous materials. We assume that a sample is composed of an association of several phases i with characteristics ε_i, σ_i. If a step function of voltage is applied to this system, the initial potential distribution is that which corresponds exclusively to the spacial distribution of the permittivities ε_i, regardless of the conductivities σ_i. On the other hand, the steady potential distribution is that which corresponds exclusively to the spacial distribution of the resistivities (assuming that charge accumulation at the boundaries does not perturb the field). In other words, if the capacitive distribution and the resistive distribution, mentioned above, are not the same, the system changes with time from one to the other. Thus, the "response" to an alternating voltage is a relaxation-like process which is called the Maxwell-Wagner relaxation (M.W.).

In the following, we shall consider the simplest case of M.W. relaxation: that of a double layer.

Figure 39 represents a double-layered structure with characteristics ε_1, ε_2, and σ_1 and σ_2 on the left, and the equivalent circuit on the right.

If we call C^* the complex capacity, per unit area, of the structure, we can write:

$$C^* = \frac{\varepsilon^*}{d} \tag{VII.39}$$

where ε^* is the effective, equivalent permittivity, and $d = d_1 + d_2$.

From the equivalent circuit, it is obvious that

$$\frac{1}{C^*} = \frac{1}{C_1^*} + \frac{1}{C_2^*} \tag{VII.40}$$

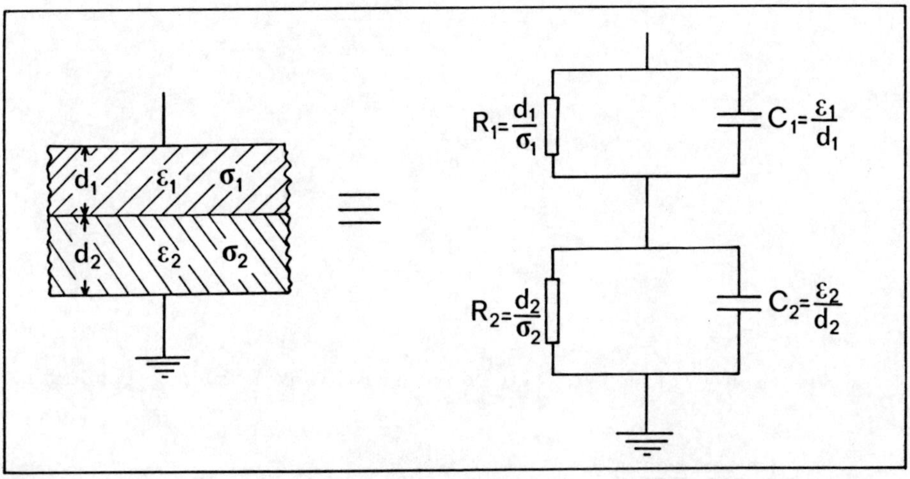

Fig.39- A double-layer capacitor and its equivalent circuit.

or

$$\frac{d}{\varepsilon^*} = \frac{d_1}{\varepsilon_1^*} + \frac{d_2}{\varepsilon_2^*} \qquad \text{(VII.41)}$$

and, coming back to the expression of the complex permittivity ε^* in terms of the real part ε and the imaginary part σ/ω , equation (VII.41) gives:

$$\varepsilon^* = d \left[\frac{d_1}{\varepsilon_1 - i\frac{\sigma_1}{\omega}} + \frac{d_2}{\varepsilon_2 - i\frac{\sigma_2}{\omega}} \right]^{-1} \qquad \text{(VII.42)}$$

After some tedious algebra, this can be written in the form:

$$\varepsilon^* = \frac{d}{\frac{d_1}{\varepsilon_1} + \frac{d_2}{\varepsilon_2}} \left[1 + \left(\frac{\tau_1 - \tau_2}{\tau_o} \right)^2 \frac{\varepsilon_o^2 \; d_1 \; d_2}{\varepsilon_1 \varepsilon_2 d_2^2 \; (1 + i\omega\tau)} \right] + \frac{1}{i\omega\tau_o} \qquad \text{(VII.43)}$$

where $\tau_1 = \varepsilon_1/\sigma_1$, $\tau_2 = \varepsilon_2/\sigma_2$,

$$\tau_o = \frac{\varepsilon_o}{d} \; \frac{d_1\sigma_2 + d_2\sigma_1}{\sigma_1 \; \sigma_2} \; , \quad \text{and} \quad \tau = \frac{d_1\varepsilon_2 + d_2\varepsilon_1}{d_1 \, {}_2 + d_2\sigma_1}$$

This in turn can be written as:

$$\varepsilon^*(\omega) = \varepsilon_\infty + \frac{\varepsilon_s - \varepsilon_\infty}{1 + i\omega\tau} - i\,\frac{\varepsilon_o}{\omega\tau_o} \qquad \text{(VII.44)}$$

where

$$\varepsilon_s = \frac{d\sigma_1\sigma_2\,(\varepsilon_1 d_2 + \varepsilon_2 d_1)}{(\sigma_1 d_2 + \sigma_2 d_1)^2} \qquad \text{(VII.45)}$$

and

$$\varepsilon_\infty = \frac{d\,\varepsilon_1\,\varepsilon_2}{\varepsilon_1 d_2 + \varepsilon_2 d_1} \qquad \text{(VII.46)}$$

Equation (VII.44) is the Debye relation including conductivity.

Particular cases

1 - One of the constituents is perfectly insulating ($\sigma_1 = 0$). This implies that $\tau_1 = \tau_o = \tau$, and equation (VII.44) reduces to a simple Debye relation.

2 - If $\tau_1 = \tau_2$, or $\varepsilon_1\sigma_2 = \varepsilon_2\sigma_1$, the relaxation term disappears and

$$\varepsilon^*(\omega) = \frac{d}{\frac{d_1}{\varepsilon_1} + \frac{d_2}{\varepsilon_2}} - \frac{i\,\varepsilon_o}{\omega\tau_o} \qquad \text{(VII.47)}$$

This is shown in Figure 40.

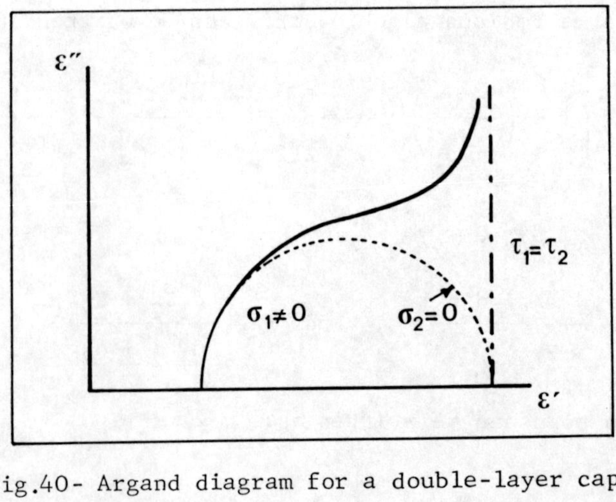

Fig.40- Argand diagram for a double-layer capacitor.

Problem

Study of a double-layer capacitor

A capacitor consists of two superposed layers, thickness d_1 and d_2, of non-polar materials between plane parallel electrodes. Each of these materials has a permittivity (ε_1 and ε_2) and a conductivity (σ_1 and σ_2) independent of the electric field. The thickness/area ratio is such that edge effects can be neglected.

One of the electrodes (A) is grounded, the other (B) is connected to a voltage source of low impedance ($V_B = V$). The interface between the layers 1 and 2 is at voltage v.

1. Draw the equivalent circuit of the unit area of the capacitor, and find the expressions for the resistances R_1, R_2 and the capacitances C_1 and C_2.

2. The electrode B, initially grounded, is suddenly raised to a voltage V. Find the magnitude of v immediately after the application of V on B, and at $t = \infty$. What must be the relationship between the characteristics of materials 1 and 2 in order that $v(\infty) = v(o)$? What is $v(t)$ in this case ?

3. One of the layers is assumed to be air. This is perfectly insulating provided that the field in the layer remains smaller than a critical value E_c, and perfectly conducting if $E \gg E_c$. (The breakdown field in air is only about 1% of that in the other layer.) Can the condition of question 2 be satisfied ? How does $v(t)$ vary in this case ?

4. V is now of the form $V_o \exp i\omega t$. Draw the Argand diagram $\varepsilon''(\varepsilon')$ for the effective complex permittivity of the composite material, and note the particular values of ε' and ε'' on the diagram.

5. The magnitude of V_o is limited by the maximum field E_c in the air layer. How does the maximum value of V_o vary with ω ?

6. A step function of voltage is applied to increase the field in the air layer from an initial value below E_c to a final value above E_c.

As soon as the field in the air layer reaches E_c, a discharge shorts the air gap. If the deionizing time is instantaneous, the discharge stops immediately and the field across the air layer starts rising again until it reaches the size E_c for a second time. A second discharge occurs, and the process repeats itself periodically.

Calculate the variation of the field across the air layer, and the period of recurrence.

7. Finally, the applied voltage is again sinusoidal, with V_o large enough to produce breakdown in the air gap. Discuss the variation of v in this case.

Case of an heterogeneous material

The above treatment can be extended to a dispersion of slightly conducting spherical particles of radius a, permittivity ε_i and conductivity σ_i in a matrix of permittivity ε_e.

In order to calculate the effective permittivity of the dispersion, consider a sphere of radius $R \gg a$ containing n uniformly distributed spherulites of permittivity ε_i, embedded in an external continuum of permittivity ε_e (that of the unfilled matrix), in a field E.

From classical electrostatics (cf. Section V.3), the potential at point $M(r,\theta)$ outside a sphere of radius R and of permittivity ε in a field E_0 (Fig.41) is:

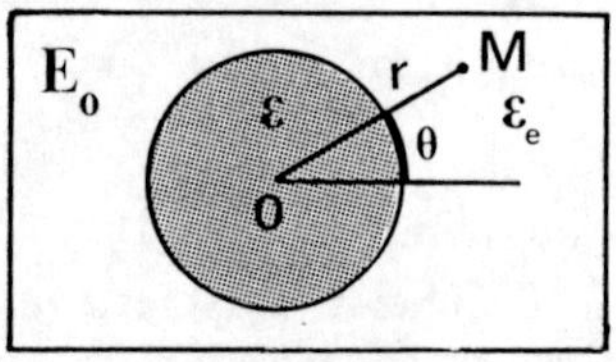

Fig. 41

$$V_e = -\left(1 - \frac{R^3}{r^3}\frac{\varepsilon - \varepsilon_e}{\varepsilon + 2\varepsilon_e}\right)E_o r\cos\theta \qquad \text{(VII.48)}$$

If the spherulites in the sphere of radius R now coalesce in its centre, to form a concentric sphere of radius $\sqrt[3]{na}$, the potential outside the large sphere is unaffected, but it is now given by:

$$V_e = -\left(1 - \frac{na^3}{r^3}\frac{\varepsilon_i - \varepsilon_e}{\varepsilon_i + 2\varepsilon_e}\right)E_o r\cos\theta \qquad \text{(VII.49)}$$

By solution of equations (VII.48) and (VII.49), we have:

$$\frac{\varepsilon - \varepsilon_e}{\varepsilon + 2\varepsilon_e} = \frac{na^3}{R^3}\,\frac{\varepsilon_i - \varepsilon_e}{\varepsilon_i + 2\varepsilon_e} \qquad \text{(VII.50)}$$

Provided that $\frac{na^3}{R^3} \ll 1$ (case of a dilute dispersion), the value of ε from eqn.(VII.50) is:

$$\varepsilon \simeq \varepsilon_e \left(1 + \frac{3na^3}{R^3}\,\frac{\varepsilon_i - \varepsilon_e}{\varepsilon_i + 2\varepsilon_e}\right) \tag{VII.51}$$

At high frequencies, $\varepsilon^* = \varepsilon_\infty$ as given by equation (VII.51).

On the other hand, at very low frequencies, the conductivity σ_i of the particles plays a major role. The particles are equipotentials, which implies that $\varepsilon_i \rightarrow \infty$, and equation (VII.51) becomes:

$$\varepsilon_s = \varepsilon_e \left(1 + \frac{3na^3}{R^3}\right) \tag{VII.52}$$

In an alternating field, the system relaxes with a time constant $\tau = (\varepsilon_i + 2\varepsilon_e)/\sigma_i$, and its complex permittivity is given by the Debye relation:

$$\varepsilon^* = \varepsilon_\infty + \frac{\varepsilon_s - \varepsilon_\infty}{1 + i\omega\tau}$$

This treatment for spherical particles has been extended in turn by Sillars to ellipsoids of revolution of axes a and b . The results are similar to those given above, except that the denominator $\varepsilon_i + 2\varepsilon_e$ of eqn.(VII.51) is replaced by $\varepsilon_i + (m-1)\varepsilon_e$, where m is related to the eccentricity $e = (a-b)/a$ of the ellipsoid by:

$$m = e^2 \left[1 - \sqrt{1-e^2}\,\frac{\sin^{-1} e}{e}\right]^{-1} \tag{VII.53}$$

For $e = 0$ (case of a sphere), equation (VII.53) gives $m = 3$, in agreement with the above results on spherical particles. If a dispersion of elongated particles is investigated, proper fitting of the experimental data with Sillar's results can be used to determine the shape of the dispersed particles. This technique has been successfully applied to biological systems.

VII.8. Dipole relaxation of defects in crystal lattices

The model

We consider, as an example, the LiF lattice of Section IV.2, containing N substitutional Mg ions per unit volume. Let $n_1(t)$, $n_2(t)$ and $n_3(t)$ be the concentrations of dipoles of types 1, 2 and 3, respectively. Obviously,

$$n_1(t) + n_2(t) + n_3(t) = N \qquad \text{(VII.54)}$$

The quantities $n_1(t)$, $n_2(t)$ and $n_3(t)$ fluctuate about their instantaneous ensemble averages $\langle n_1(t)\rangle$, $\langle n_2(t)\rangle$ and $\langle n_3(t)\rangle$, because of thermal motion.

In the absence of an applied field, $n_1(t)$, $n_2(t)$ and $n_3(t)$ are different at a given instant, but

$$\langle n_1(t)\rangle = \langle n_2(t)\rangle = \langle n_3(t)\rangle = \frac{N}{3} \qquad \text{(VII.55)}$$

In the remaining paragraphs of this section, n_i will be used for $\langle n_i(t)\rangle$ to simplify notation, but the meaning of this notation should be kept in mind.

If p_{ij} is the probability of a transition between a configuration i to a configuration j $(i,j = 1,2,3)$, the dynamical equations for the Mg-doped LiF, using the notations introduced Section IV.2, are:

$$\left.\begin{aligned} \dot{n} &= -(p_{12} + p_{13})n_1 + p_{21}\, n_2 + p_{31}\, n_3 \\ \dot{n}_2 &= p_{12}\, n_1 - (p_{21} + p_{23})\, n_2 + p_{32}\, n_3 \\ \dot{n}_3 &= p_{13}\, n_1 + p_{23}\, n_2 - (p_{31} + p_{32})\, n_3 \end{aligned}\right\} \qquad \text{(VII.56)}$$

The coefficients p depend on the applied field, but their number can be reduced from 6 to 4 if we keep in mind that only two types of transitions may occur in the absence of an applied field, i.e.:

1. Hopping of a $\boxed{Li}^-$ between adjacent sites within a unit cell, with a probability

$$p_{12}(0) = p_{21}(0) = p_{32}(0) = p_{23}(0) = \varpi_1$$

2. Hopping of a $\boxed{Li}^-$ from one cell to the next, with a probability

$$p_{13}(0) = p_{31}(0) = \varpi_2$$

In the presence of an applied field, the probabilities $p(0)$ becomes:

$$p_{12} = p_{23} = \varpi_1 e^{-y}$$

$$p_{21} = p_{32} = \varpi_1 e^{-y}$$

$$p_{13} = \varpi_2 e^{-2y}$$

$$p_{31} = \varpi_2 e^{2y}$$

with $y = \frac{aeE}{2kT}$, as in eqn.(IV.14).

Introducing these values in equations (VII.56) gives:

$$\dot{n}_1 = \varpi_1(n_2e^y - n_1e^{-y}) + \varpi_2(n_3e^{2y} - n_1e^{-2y}) \qquad \text{(VII.57a)}$$

$$\dot{n}_2 = \varpi_1\left[n_1e^{-y} - 2n_2\text{ch}y + n_3e^y\right] \qquad \text{(VII.57b)}$$

$$\dot{n}_3 = \varpi_1(n_2e^{-y} - n_3e^y) - \varpi_2(n_3e^{2y} - n_1e^{-2y}) \qquad \text{(VII.57c)}$$

By adding these three equations, we have

$$\dot{n}_1 + \dot{n}_2 + \dot{n}_3 = 0, \qquad \text{(VII.57d)}$$

which can be obtained directly by differentiation of eqn.(VII.54).

Consequently, these equations are not redundant and determine the three unknown n_1, n_2 and n_3.

The linear combination $\dot{n}_1e^{-y} - \dot{n}_3e^y$ gives:

$$\dot{n}_1e^{-y} - \dot{n}_3e^y = (\varpi_1 + 2\varpi_2\cosh y)(n_3e^{2y} - n_1e^{-2y}) \qquad \text{(VII.58)}$$

Since y is usually very small, the exponentials can be replaced by their first-order expansions, and eqn.(VII.58) becomes:

$$\dot{n}_1 - \dot{n}_3 + y\,\dot{n}_2 = (\varpi_1 + 2\varpi_2)\left[n_3 - n_1 + 2y(N - n_2)\right] \qquad \text{(VII.59)}$$

A particularly simple case is that of a small field oscillating about zero, so that n_2 stays constant:

$$n_2 = \frac{N}{3} \qquad \text{(VII.60)}$$

$$\dot{n}_1 + \dot{n}_3 = \dot{n} = 0 \qquad \text{(VII.61)}$$

Now, if we let $\varpi_1 + 2\varpi_2 = 1/\tau$ and $n_1 - n_3 = n$, eqn.(VII.59) becomes:

$$\tau\dot{n} + n - \frac{4}{3}\,yN = 0 \qquad \text{(VII.62)}$$

We now assume that the field E varies as $E_o \exp i\omega t$, so that $y = y_o \exp i\omega t$. In the steady state, all the variable quantities oscillate with the same angular frequency ω, and we can write, by symmetry:

$$n_1 = \frac{N}{3} + \nu \exp i\omega t \qquad \text{(VII.63a)}$$

$$n_3 = \frac{N}{3} - \nu \exp i\omega t \qquad \text{(VII.63b)}$$

where ν is small. From eqns.(VII.63a) and (VII.63b),

$$n_1 - n_3 = n = 2\nu \exp i\omega t \qquad \text{(VII.64)}$$

Introducing this value of n into eqn.(VII.62) gives

$$2i\omega\nu\tau + 2\nu - \frac{4}{3}\,y_oN = 0 \qquad \text{(VII.65)}$$

from which ν, and therefore n, is obtained. Using the same averaging procedure as in Section IV.2, the polarization due to the dipole defects is:

$$P_d = \frac{Na^2e^2}{6kT(1 + i\omega\tau)}\,E \qquad \text{(VII.66)}$$

Using now the value of μ ($\mu = ae/\sqrt{2}$), the complex permittivity of the lattice takes the familiar form

$$\varepsilon^* = \varepsilon + \frac{N\mu^2}{3kT(1 + i\omega\tau)} \qquad \text{(VII.67)}$$

where ε accounts for the ionic and electronic components of the frequency independent permittivity.

We recognize eqn.(VII.67) as the equation of a Debye relaxation process. Indeed, magnesium and calcium doped lithium fluorides have a characteristic Debye relaxation diagram, from which the dopant concentration and the relaxation time τ can be deduced.

Many other crystals (and even polycrystalline samples) containing mobile lattice defects present similar relaxation spectra. The study of dielectric relaxation is a powerful tool for the study of lattice defects. For instance, a good part of our present understanding of the structure of colour centres results from dielectric relaxation spectra. However, other tools such as nuclear magnetic resonance, optical and Raman spectroscopy can be used efficiently in conjunction with dielectric spectroscopy.

VII.9. Space-charge polarization and relaxation

The problem of a relaxing space-charge is of major practical importance. It is related to the accumulation of charges in the vicinity of blocking electrodes, i.e. electrodes which are unable to discharge the ions arriving on them.

The ions experience the combined influence of the field which tends to accumulate the charges on the electrodes, and thermal diffusion which tends to oppose this accumulation.

To illustrate this process, we note that in steady-state equilibrium under a constant applied field, a sheet of material is macroscopically polarized. It contains a charge density $\rho(x)$ which depends only on the depth x, and is therefore equivalent to a large dipole

$$\mu = \int x\, \rho(x)\, dx$$

(configuration (a) of Figure 42).

If the direction of polarization is now reversed, the space-charge distribution slowly evolves towards a new steady state (b) with the same symmetry with respect to the middle plane as (a), and the sample becomes equivalent to an opposite dipole $(-\mu)$. The sluggish dipole reversal is a relaxation, which we shall now investigate mathematically, using a few simplifying assumptions.

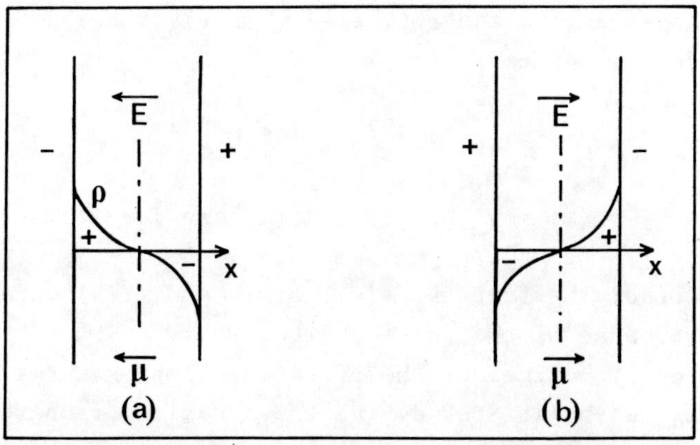

Fig.42- Model for space-charge relaxation.

Simplifying assumptions

(a) Only one type of charge carrier can move (positive carriers, for instance, to avoid errors with the signs) in a continuum of permittivity ε .

(b) The applied field is so weak that the system remains linear, and the space and time variables can be separated.

Charges of both signs are usually mobile, with different mobilities. They recombine to some extent on the electrodes and between themselves, and carriers are thermally generated in the bulk. Accounting for these facts changes the results quantitatively but does not alter the main steps of the following derivation which is given here as an example.

Let n_o be the uniform concentration of mobile charges (neutralized by the same density of fixed negative charges) in the absence of an applied field, and μ be the mobility of these positive charges.

The sheet of material located between the planes $x = -d$ and $x = +d$ (Fig.42) experiences an alternating field of the form:

$$E_a = \mathcal{E}_a \exp i\omega t \tag{VII.68}$$

where $\mathcal{E}_a$ satisfies assumption (b) above.

Under the combined action of the applied and diffusion fields, the charge concentration at x is only slightly different from n_o , so that, in the a.c. steady state, the difference $(n - n_o)$ oscillates at the same frequency as the applied field, and we can write, as before:

$$n(x,t) = n_o + \nu(x) \exp i\omega t \tag{VII.69}$$

In the same way, the potential $V(x,t)$ and the field $E(x,t)$ take the respective forms:

$$V(x,t) = \varphi(x) \exp i\omega t \tag{VII.70}$$

and

$$E(x,t) = \mathcal{E}(x) \exp i\omega t \tag{VII.71}$$

Of course, the factors $\nu(x)$, $\varphi(x)$ and $\mathcal{E}(x)$ are complex quantities, because of the phase shifts.

In order to calculate the effective complex permittivity ε^* of the material, we shall first derive the equation of charge conservation. This equation reduces to a simple second-order differential equation for

$\nu(x)$, which can be solved. From $\nu(x)$, the charge density, and subsequently the amplitude of the a.c. polarization $P(\omega)$ can be obtained by calculating the effective dipole moment per unit volume. Finally, it may be shown from the theorems mentioned above that $\varepsilon^* = \varepsilon + P(\omega)/\mathcal{E}_a$.

Another less direct but perhaps more instructive method is given as a problem.

Derivation of $\nu(x)$

The "particle" current density at x and t (i.e. the number of particles crossing a unit area of the plane x at time t) is the sum of the conduction current $\mu n \vec{E}$ and the diffusion current $-D\vec{\nabla} n$:

$$j(x,t) = -\mu n \frac{\partial V}{\partial x} - D\frac{\partial n}{\partial x} \qquad \text{(VII.72)}$$

and the equation for the conservation of the charge

$$\frac{\partial n}{\partial t} = -\frac{\partial j}{\partial x} \qquad \text{(VII.73)}$$

becomes

$$\frac{\partial n}{\partial t} = \mu \frac{\partial}{\partial x}\left(n\frac{\partial V}{\partial x}\right) + D\frac{\partial^2 n}{\partial x^2} \qquad \text{(VII.74)}$$

Substituting (VII.69) and (VII.70):

$$i\omega\nu = \mu n_o \frac{d^2\varphi}{dx^2} + D\frac{d^2\nu}{dx^2} + \mu\frac{d}{dx}\left(\nu\frac{d\varphi}{dx}\right)e^{i\omega t} \qquad \text{(VII.75)}$$

Since ν is a small quantity, the magnitude of the third term of the right hand side of eqn.(VII.75) is small compared with that of the first two terms. Poisson's equation gives:

$$\frac{d^2\varphi}{dx^2} = -\frac{e}{\varepsilon}\nu , \qquad \text{(VII.76)}$$

so that (VII.75) becomes finally

$$\left(i\omega + \frac{\mu e n_o}{\varepsilon}\right)\nu = D\frac{d^2\nu}{dx^2} \qquad \text{(VII.77)}$$

but

$$\frac{n_o \mu' e}{\varepsilon} = \frac{\sigma}{\varepsilon} = \frac{1}{\tau} ,$$

where τ is the relaxation time of the conductivity. Therefore, (VII.77) becomes

$$\frac{d^2\nu}{dx^2} = \frac{1 + i\omega\tau}{D\tau}\,\nu \tag{VII.78}$$

Using now the reduced variable

$$X = \sqrt{\frac{1 + i\omega\tau}{D\tau}}\, x = \frac{Z}{\lambda}\, x \tag{VII.79}$$

($Z = \sqrt{1 + i\omega\tau}$ and $\lambda = \sqrt{D\tau}$ is the Debye length), equation (VII.77) now takes the simple form:

$$\frac{d^2\nu}{dX^2} = \nu \tag{VII.80}$$

and its general solution

$$\nu = \nu_1\, e^{X} + \nu_2\, e^{-X} \tag{VII.81}$$

must obey the boundary condition

$$\int_{-d}^{d} \nu(x)\, dx = 0$$

This implies that $\nu_2 = -\nu_1$, so that the actual solution is

$$\nu = \nu_1 \sinh X = \nu_1 \sinh\left(\frac{Z}{\lambda}\, x\right) \tag{VII.82}$$

From ν, Poisson's equation gives

$$\mathcal{E}(x) = \frac{e}{\varepsilon}\int \nu dx = \frac{e\nu_1}{\varepsilon}\,\frac{\lambda}{Z}\cosh\left(\frac{Z}{\lambda}\, x\right) + \mathcal{E}_o \tag{VII.83}$$

with the boundary condition on the voltage:

$\int_{-d}^{+d} E(x)\, dx = 2\,\mathcal{E}_a\, d$, which becomes here

$$\frac{e\nu_1}{\varepsilon}\left(\frac{\lambda}{Z}\right)^2 \sinh\left(\frac{Z}{\lambda}\, d\right) + \mathcal{E}_o = \mathcal{E}_a d \tag{VII.84}$$

Using the reduced variable X to simplify notation, and letting $\frac{Zd}{\lambda} = \sqrt{1 + i\omega\tau}\,\frac{d}{\lambda} = Y$

$$\mathcal{E}(X) = \mathcal{E}_a + \nu_1 \frac{de}{\varepsilon X_m} (\cosh X - \frac{\sinh Y}{Y}) \qquad \text{(VII.85)}$$

Finally, the value of the constant ν_1 is obtained from the requirement that the particle current at $X = X_m$ is zero, since the electrodes are assumed to be blocking:

$$n \mu \mathcal{E}(Y) - D \frac{Z}{\lambda} \left(\frac{d\nu}{dX}\right)_Y = 0 \, , \qquad \text{(VII.86)}$$

from which

$$\mathcal{E}(Y) = \mathcal{E}_a + \nu_1 \frac{de}{\varepsilon Y} (\cosh X_m - \frac{\sin Y}{Y}) = D \frac{Z\nu_1}{n\mu} \cosh Y \qquad \text{(VII.87)}$$

Using Einstein's relation $(D/\mu = kT/e)$ and keeping in mind that $n \simeq n_o$, ν_1 can be deduced from the above equation, after some algebraic manipulation:

$$\nu_1 = \frac{\varepsilon Z}{e\lambda} \mathcal{E}_a (i\omega\tau \cosh Y + \frac{\sin Y}{Y})^{-1} \qquad \text{(VII.88)}$$

so that

$$\nu(X) = \frac{\varepsilon Z}{e\lambda} \mathcal{E}_a \sinh X (i\omega\tau \cos Y + \frac{\sin Y}{Y})^{-1} \qquad \text{(VII.89)}$$

To calculate $\mathcal{E}(\omega)$, we can use the first method mentioned above; we calculate the polarization or dipole moment per unit volume:

$$P(\omega) = \frac{e}{2d} \int_{-d}^{d} x \, \nu(x) \, dx = \left(\frac{\lambda}{Z}\right)^2 \frac{e}{d} \int_{o}^{Y} X \nu(X) dX$$

or

$$P(\omega) = \frac{\lambda\varepsilon}{Zd} \frac{\mathcal{E}_a}{i\omega\tau \cosh Y + \sinh Y/Y} \int_{o}^{Y} X \cosh X \, dX \qquad \text{(VII.90)}$$

After integration, $P(\omega)$ becomes

$$P(\omega) = \frac{\lambda\varepsilon}{Zd} \mathcal{E}_a \frac{Y\cosh Y - \sinh Y}{i\omega\tau \cosh Y + \sinh Y/Y} \qquad \text{(VII.91)}$$

and, using the definition $\varepsilon^* : \varepsilon^* = \varepsilon + \frac{P}{\mathcal{E}_a}$,

$$\boxed{\varepsilon^* = \varepsilon \frac{1 + i\omega\tau}{i\omega\tau + \tanh Y/Y}} \qquad \text{(VII.92)}$$

This equation shows that,

(a) for $\omega = 0$, $\varepsilon^* = \varepsilon_s = \varepsilon \dfrac{d/\lambda}{\tanh(d/\lambda)} = \varepsilon\,\delta \,\mathrm{cotanh}\,\delta$

with $\delta = d/\lambda$

(b) for $\omega \to \infty$, $\varepsilon^* \to \varepsilon_\infty = \varepsilon$

Unlike the Debye relation, the expression (VII.92) cannot be easily separated into its real and imaginary parts. The calculation, which is possible but rather tedious, is proposed as an exercise to those of sound nervous disposition ! Computers can perform this sort of calculation very well, for all possible values of $\omega\tau$. Typical results are shown in Figure 43 (a) and (b) which represent $x = \varepsilon'/\varepsilon$ and $y = \varepsilon''/\varepsilon$ respectively, as a function of $\omega\tau$, for various values of the parameter $\delta = d/\lambda$.

Argand diagrams can also be constructed, and they are found to be almost exactly semi-circular provided that $d \gg \lambda$, and slightly flattened if $d \ll \lambda$. Fig. 44 shows two reduced diagrams of the dimensionless quantity $\dfrac{\varepsilon^* - \varepsilon_\infty}{\varepsilon_s - \varepsilon_\infty}$, which varies from 1 to 0 as $\omega\tau$ varies from 0 to ∞. The diagram is almost exactly a semicircle in the first case; it is slightly flattened in the second.

We shall now give, as a problem, an alternative derivation of equation (VII.92) for the effective complex permittivity.

Problem

Another method to derive $\varepsilon(\omega)$

1. From the expression of $v(X)$, deduce that the amplitude of the field $\mathcal{E}(X)$ is

$$\mathcal{E}(X) = \mathcal{E}_a \frac{\cosh X + i\omega\tau \cosh\delta}{\sinh\delta/\delta + i\omega\tau\cosh\delta}, \quad \text{with } \delta = \frac{d}{\lambda}$$

2. Write the complete expression for the "local" current density, including the conduction and the diffusion currents, as a function of x and t . Deduce from this the corresponding contribution of the total current in the external circuit by using the relation

$$J(t) = \frac{1}{2d}\int_{-d}^{+d} J(x,t)\,dx$$

which should be derived.

3. Write the displacement current in the a.c. steady state. By adding this displacement current to the contribution from the particle current obtained in question 2, write down the total current in the external circuit.

4. To find ε^* , compare the current obtained in question 3 to that which flows under the same conditions, in a sample of complex permittivity ε^* . Equation (VII.92) should be obtained.

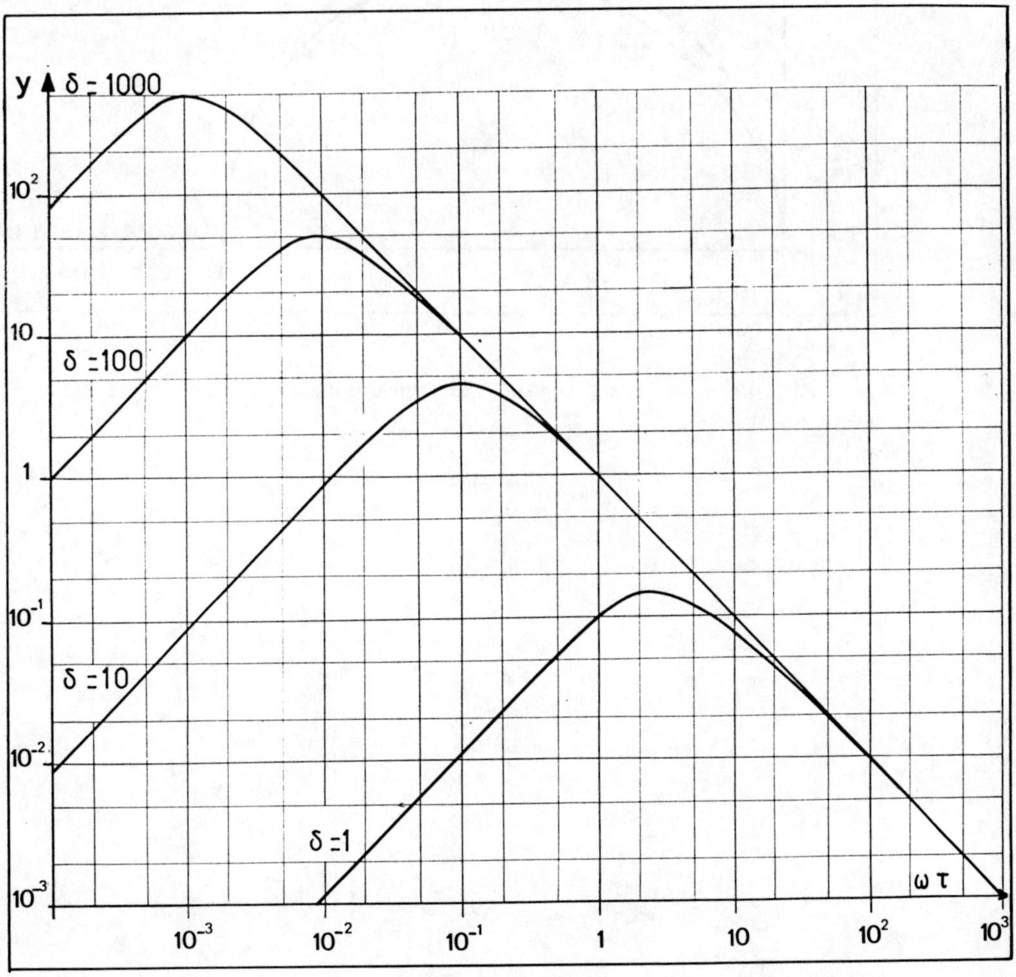

Fig.43 - Dissipation spectrum for sphace-chargė relaxation.

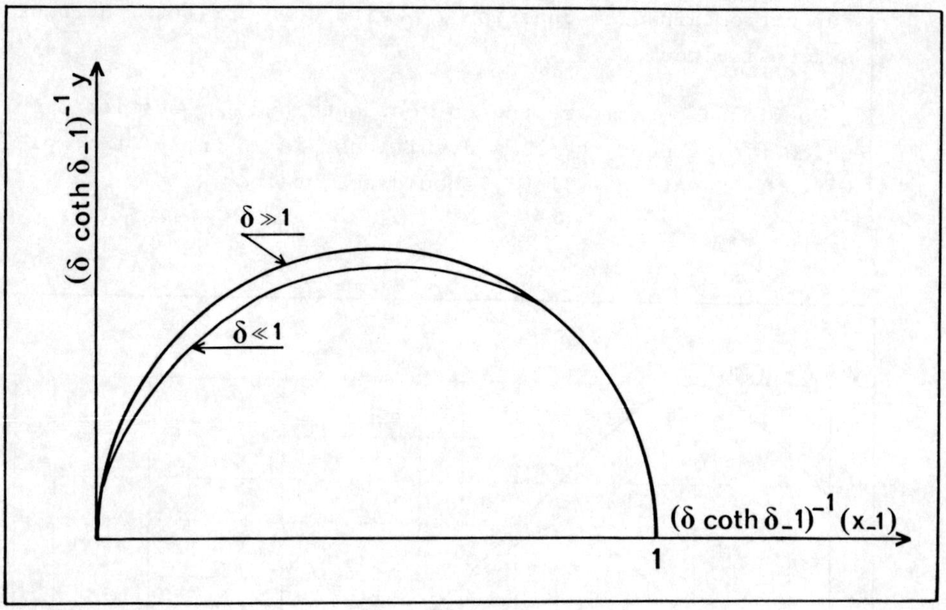

Fig.44 - Reduced Argand Diagrams from equation (VII.92).

VII.10. Recent work - Many-body interpretation

The advent of statistical mechanics, and in particular the general formulation of Kubo, has generated a brilliant development of the microscopic theories of relaxation in the liquid phase. The names of Glarum, Cole, Williams, Davies and Brot among others, are familiar to anyone specialising in this field, and, although the development of these theories is beyond the scope of this book, the main references are given in the bibliography.

In solids, typical loss peaks are broader than the Debye peak ; they may be asymmetric and their frequency is often too low for natural oscillations of a one-particle system, e.g. dipole.

Distributed relaxation times may account for these departures from the ideal response but do not seem physically satisfactory.

A recent approach of A. Jonscher, starting from the Ising model of nearest-neighbour interactions, predicts loss peaks with a broader width, an asymmetric shape and a lower frequency of maximum than the "one particle" loss peak. This seems well suited to the case of polymeric and glassy materials, and may also apply to the case of ferroelectrics , which all show this type of response.

Once the peak width exceeds 3 to 4 decades, as many low-temperature peaks in polymers do, the above interpretation is no longer suitable, and other mechanisms, still involving many-body interactions, have to be considered.

A limiting situation arises in some materials in which a low loss is practically independent of frequency over several decades. This behaviour departs so widely from the Debye response that it requires a completely different approach, such as many-body systems can provide.

Although the present understanding of these processes is not yet sufficient to permit quantitative predictions, it now appears that the dielectric response of solids may offer a valuable insight into the mechanisms of many-body interactions.

As a conclusion of this chapter, dielectric spectroscopy is an excellent diagnostic for dilute dipole systems. It first gives the difference $(\varepsilon_s - \varepsilon_\infty)$, equal or close to $N\mu^2/3kT$, and hence the value of the dipole moment μ if their concentration is known. The upper part of the Argand diagram yields further information on the average relaxation time, and its distributions, related to the morphology and dynamical behaviour of the constituent molecules. Dielectric spectroscopy of condensed dipole systems is a wide-open, promising field.

VIII. RESONANCES

It was explained earlier, and is well known to anyone with an elementary knowledge of classical mechanics, that a resonance of a vibrating system can be produced by an excitation that oscillates at a frequency close to the natural frequency of the system. Unlike a relaxation, where the restoring action is a diffusion force of thermodynamic origin, the restoring force is now a local, pseudo-elastic force like that of a spring, totally independent of the surroundings.

If the system is frictionless, which is never rigorously the case in practice, resonance occurs only when the excitation has exactly the same frequency as the natural frequency of the system, and the amplitude of the oscillation increases without limit while the oscillating force is acting, so that the steady state oscillation is of infinite amplitude. After the excitation is removed, there is a cyclic exchange of potential and kinetic energy, but the total energy (potential + kinetic) remains constant.

If friction exists, there is a steady state oscillation of finite amplitude in a narrow, but finite frequency range of the applied field the oscillating charges polarize the system, causing a phase shift with respect to the field. This is the case with oscillating charges which, according to electromagnetic theory, radiate energy when accelerated, and consequently undergo "radiation damping". Consequently, the complex permittivity of the system can be obtained, as a function of the frequency, by comparing the amplitude of the oscillating polarization with that of the field.

Ionic and molecular vibrations should strictly be treated by quantum mechanics, but, as mentioned previously, the correspondence principle permits quantum systems of high quantum numbers (i.e. with large masses and vibrational energy) to be treated by classical mechanics. This is shown in the next section.

VIII.1. The linear oscillator model

Consider a material composed of identical linear oscillators of reduced mass m, charge e, elastic constant β and friction coefficient f. Let E' be the local field acting on each oscillator (we assume that the wave-length of the electromagnetic fields is large with respect to the system).

The equation of motion of each oscillator in the field direction x is:

$$m\ddot{x} = eE' - \beta x - f\dot{x} \quad \text{(VIII.1)}$$

(applied-elastic-friction)

This equation is of the form:

$$\ddot{x} + 2\gamma\dot{x} + \omega_o^2 = \frac{e}{m}E' \quad \text{(VIII.2)}$$

where $\gamma = \frac{f}{2m}$ and $\omega_o = \sqrt{\frac{\beta}{m}}$ = angular frequency of the undamped oscillator.

If N is the concentration of the oscillators, the polarization of the system is:

$$P = N e x \quad \text{(VIII.3)}$$

The local field E' can be related to the polarization and the applied field by the Lorentz relation:

$$E' = E + \frac{P}{3\varepsilon_o} \quad \text{(VIII.4)}$$

Introducing eqns.(VIII.3) and (VIII.4) into (VIII.2) gives

$$\ddot{P} + 2\gamma\dot{P} + \left(\omega_o^2 - \frac{Ne^2}{3m\varepsilon_o}\right)P = \frac{Ne^2}{m}E \quad \text{(VIII.5)}$$

If the applied field E oscillates with an angular frequency ω, $E = E_o \exp i\omega t$. The polarization also oscillates with the same frequency, once the steady state is established (i.e. after the transient has died away). However, there is a phase shift φ between E and P, so that:

$$P = P_o \exp i(\omega t + \varphi) \quad \text{(VIII.6)}$$

Using the above expression for P in eqn. (VII.5) gives:

$$\left(-\omega^2 + 2i\gamma\omega + \omega_o^2 - \frac{Ne^2}{3m\varepsilon_o}\right)P = \frac{Ne^2}{m}E \quad \text{(VIII.7)}$$

From the magnitudes of the steady-state polarization, the complex permittivity is obtained:

$$\varepsilon^* = \varepsilon_o + \frac{P}{E} = \varepsilon_o + \frac{Ne^2/m}{\omega_o'^2 - \omega^2 + 2i\gamma\omega} \quad \text{(VIII.8)}$$

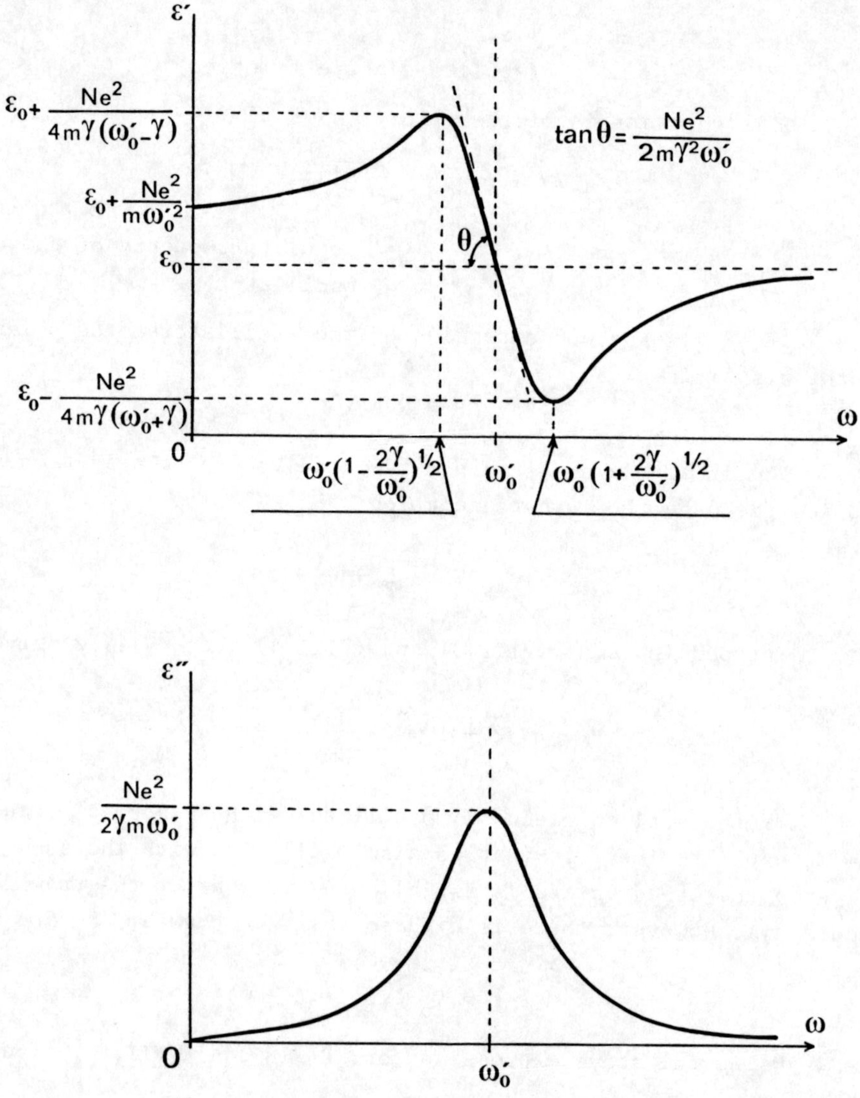

Fig. 45 - $\varepsilon'(\omega)$ and $\varepsilon''(\omega)$ for damped harmonic oscillators.

where we have used the notation

$$\omega_o'^2 = \omega_o^2 - \frac{N e^2}{3m\varepsilon_o}$$

From the expression (VIII.8) for the complex permittivity, we can extract the real and imaginary parts of ε^* :

$$\varepsilon'(\omega) = \varepsilon_o + \frac{Ne^2}{m} \frac{\omega_o'^2-\omega^2}{(\omega_o'^2 - \omega^2)^2 + 4\gamma^2\omega^2} \tag{VIII.9}$$

$$\varepsilon''(\omega) = \frac{Ne^2}{m} \frac{2\gamma\omega}{(\omega_o'^2 - \omega^2)^2 + 4\gamma^2\omega^2} \tag{VIII.10}$$

and the phase angle $\tan^{-1}(\varepsilon''/\varepsilon')$.

These quantities have been plotted in Figure 45 as a function of frequency. All the characteristic quantities are marked on the graph, where it can be seen that:

(a) the maximum of ε'', of amplitude close to $Ne^2/2\gamma m\omega_o'$, occurs for an angular frequency slightly lower than ω_o'.

(b) the low frequency component of the permittivity, $Ne^2/m\omega_o'^2$, is independent of the damping coefficient γ .

VIII.2. The unidimensional polar lattice

We treat here the simplest model of a polar crystal, a unidimensional dipole lattice consisting of an alternate chain of two kinds of ions (1 and 2) of masses m_1 and m_2 and charges (+q) and (-q) respectively. The reader is strongly advised to work out problem 4 of Section III.3 before studying this section.

The ions are regarded as oscillators connected to their neighbours by elastic forces of string constant β . They are numbered according to their parity. For instance, the ions of type 1, which are black on the figure, are pair, and the ions of type 2, which are white on the drawing, are odd. The average distance between neighbour ions is **a**.

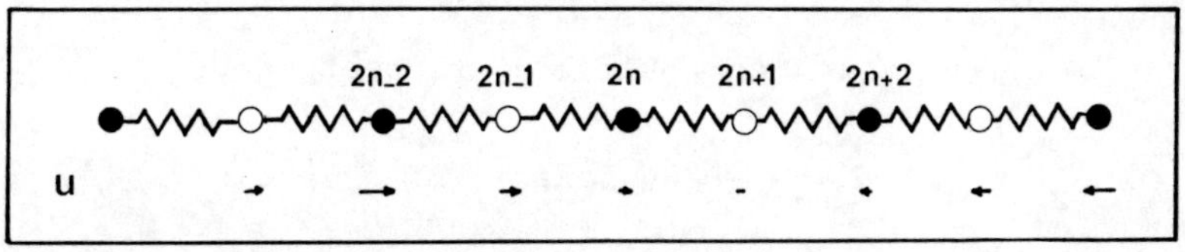

Fig.46 - Schematic representation of a diatomic linear chain.

We first write the equations of motion for both types of ions in the absence of an applied field, assuming that:

(a) only interactions between nearest neighbours have to be accounted for,
(b) the frictional forces are negligible.

Let u_{2n} be the displacement, with respect to its average position, of the "black" ion 2n. The dynamical equation for this ion is:

$$m_1 \ddot{u}_{2n} = - \beta (u_{2n} - u_{2n-1}) - \beta(u_{2n} - u_{2n+1}) \qquad \text{(VIII.11)}$$

or

$$m_1 \ddot{u}_{2n} = \beta (u_{2n-1} + u_{2n+1} - 2u_{2n}) \qquad \text{(VIII.12)}$$

In the same way, for the odd ion (2n + 1):

$$m_2 \ddot{u}_{2n+1} = \beta (u_{2n} + u_{2n+2} - 2u_{2n+1}) \qquad \text{(VIII.13)}$$

In the steady state of oscillation, all the ions oscillate with the same angular frequency ω . The "black" ions oscillate with an amplitude u_1 , and the "white" ions oscillate with an amplitude u_2 . The oscillations are not in phase. The phase shift between the oscillation of a given atom and the oscillation of that atom which is taken as the origin of phase increases linearly with the distance. For example, if ion 0 is taken as the origin, ion 2n oscillates in accordance with

$$u_{2n} = u_1 \exp i(\omega t - 2nka) \qquad \text{(VIII.14)}$$

In other words an oscillation wave of wave-length $\lambda = 2\pi/k$ propagates with a velocity ω/k.

In just the same way, the dynamical equation for the "white" atom (2n + 1) is

$$u_{2n+1} = u_2 \exp i \left[\omega t - (2n+1)ka\right] \qquad \text{(VIII.15)}$$

From (VIII.14) and (VIII.15) respectively,

$$\ddot{u}_{2n} = - \omega^2 u_1 \exp i (\omega t - 2nka) \qquad \text{(VIII.16)}$$

$$\ddot{u}_{2n+1} = - \omega^2 u_2 \exp i \left[\omega t - (2n+1)ka\right] \qquad \text{(VIII.17)}$$

Introducing (VIII.16) and (VIII.17) in (VIII.12) and (VIII.13), we obtain:

$$u_{2n+1} - \left(2 - \frac{m_1\omega^2}{\beta} \right)u_{2n} + u_{2n-1} = 0 \qquad \text{(VIII.18)}$$

$$u_{2n+2} - \left(2 - \frac{m_2\omega^2}{\beta} \right)u_{2n+1} + u_{2n} = 0 \qquad \text{(VIII.19)}$$

and finally

$$u_2 \cos ka = \left(1 - \frac{m_1}{2\beta}\omega^2 \right)u_1 \qquad \text{(VIII.20)}$$

$$u_1 \cos ka = \left(1 - \frac{m_2}{2\beta}\omega^2 \right)u_2 \qquad \text{(VIII.21)}$$

These two equations are compatible, provided that

$$\left(1 - \frac{m_1}{2\beta}\omega^2\right)\left(1 - \frac{m_2}{2\beta}\omega^2\right) - \cos^2 ka = 0 \qquad \text{(VIII.22)}$$

$$\frac{m_1 m_2}{4\beta^2}\omega^4 - \frac{m_1+m_2}{2\beta}\omega^2 + \sin^2 ka = 0 \qquad \text{(VIII.23)}$$

The eigenfrequencies of the system may be obtained from (VIII.23):

$$\omega^2 = \frac{\beta}{\bar{m}}\left[1 \pm \left(1 - \frac{4\bar{m}^2}{m_1 m_2}\sin^2 ka\right)^{1/2}\right] \qquad \text{(VIII.24)}$$

where $\bar{m} = (1/m_0 + 1/m_2)^{-1}$ is the reduced mass.

If ka is very small with respect to 2π, the wave-length of the vibration is much larger than a and the atoms oscillate practically in phase. The eigenfrequencies corresponding respectively to the + and - signs of eqn.(VIII.24) are

$$\omega'^2 = \frac{2\beta}{\bar{m}}\left(1 - \frac{\bar{m}^2}{m_1 m_2}k^2a^2 + \ldots\right) \qquad \text{(VIII.25)}$$

$$\omega''^2 = \frac{2\beta\bar{m}}{m_1 m_2}k^2a^2\left(1 - \frac{k^2a^2}{3} + \ldots\right) \qquad \text{(VIII.26)}$$

The eigenfrequency ω' corresponds to the "optical" mode of vibration, in which alternate atoms oscillate in opposite directions. This is the frequency of the fundamental infrared absorption.

The eigenfrequency ω'' of the acoustic mode corresponds to the propagation of a vibration which modulates the optical vibration. It is related to the propagation of sound.

In the presence of an electric field of amplitude E and angular frequency ω , directed along the chain in the steady state, all the quantities oscillate with the same angular frequency ω . If we further assume that the local field at all the atomic sites is the same as the applied field, (an assumption which will be discussed later), equations (VIII.18) and (VIII.19) become:

$$- m_1\omega^2 u_1 = 2\beta(u_2 - u_1) + qE \tag{VIII.27}$$

$$- m_2\omega^2 u_2 = 2\beta(u_1 - u_2) - qE \tag{VIII.28}$$

Addition of eqns.(VIII.27) and (VIII.28) gives:

$$u_2 m_2 + u_1 m_1 = 0 \tag{VIII.29}$$

and by subtraction:

$$(m_2 u_2 - m_1 u_1)\omega^2 = 4\beta(u_2 - u_1) + 2qE \tag{VIII.30}$$

From eqns.(VIII.29) and (VIII.30):

$$u_1 = \frac{qE/m_1}{\omega^2 - 2\beta/\overline{m}} \tag{VIII.31}$$

$$u_2 = \frac{qE/m_2}{\omega^2 - 2\beta/\overline{m}} \tag{VIII.32}$$

From eqn.(VIII.26), $\frac{2\beta}{\overline{m}}$ is related to the angular frequency ω' of the "optic" mode.

To calculate the permittivity of the lattice, we assume that the wavelength of the vibration is large with respect to the sample size, otherwise the dipole moment per unit volume in one half wavelength of the vibration wave will cancel that of the next half wave length. This assumption implies $ka = 0$, so that

$$u_1 = \frac{qE/m_1}{\omega^2 - \omega'^2} \tag{VIII.33}$$

$$u_2 = \frac{qE/m_2}{\omega^2 - \omega'^2} \tag{VIII.34}$$

Under this condition, the dipole moment per unit volume (the polarization) is:

$$P = Nq(u_1 - u_2) = \frac{Nq^2}{\bar{m}\,(\omega'^2 - \omega^2)}\,E \qquad \text{(VIII.35)}$$

and the permittivity

$$\varepsilon(\omega) = \varepsilon_o + \frac{Nq^2}{\bar{m}\,(\omega'^2 - \omega^2)} \qquad \text{(VIII.36)}$$

increases with ω , tends to ∞ for $\omega \to \omega'$ (fundamental i.r. absorption) since the friction has been neglected, and tends to ε_o for $\omega \to \infty$. In other words, the contribution to the ionic resonances, at a frequency lower that the eigenfrequency $2\pi\omega'$, is $Nq^2/\bar{m}\omega'^2$.

As indicated above, this treatment is a great over-simplification. The neglect of the frictional term is not qualitatively serious. Accounting for it would simply smooth out the resonance band, preventing the divergence of P and ε at $\omega = \omega'$, as was shown in Fig.45 for the single linear oscillator.

The neglect of polarization in the local field, which was assumed to be equal to the applied field, is a little more questionable.

If we try to account for the contribution of the polarization to the local field by a Lorentz-type formula (which is certainly inaccurate in a linear chain), eqns (VIII.27) and (VIII.28) become respectively:

$$- m_1 u_1 \omega^2 = (2\beta - \frac{Nq^2}{3\varepsilon_o})\,(u_2 - u_1) + qE \qquad \text{(VIII.37)}$$

$$- m_2 u_2 \omega^2 = (2\beta - \frac{Nq^2}{3\varepsilon_o})\,(u_1 - u_2) - qE \qquad \text{(VIII.38)}$$

In other words, this simply changes β into $(\beta - \frac{Nq^2}{6\varepsilon})$, and hence shifts the fundamental absorption band toward the long wave-length end of the spectrum. A similar result has already been obtained in the study of the single harmonic oscillator.

Another correction which might be important in some occasions takes account of the polarizability of the ions, which has so far been neglected. In fact, the ions in the field become polarized, all the induced dipoles moments are in the direction of the applied field, and the fields created by these dipole moments contribute additively to the local field, as discussed in Section III.4.

PART 3. DISSIPATIVE EFFECTS UNDER HIGH FIELDS

IX. INSULATORS AND WIDE - GAP SEMICONDUCTORS

IX.1. Intrinsic conduction and impurity conduction

From the electronic standpoint, an insulator can be regarded as a semiconductor with a forbidden energy gap of at least 5 eV, corresponding to the energy of the optical absorption edge of the material.

Table 11 illustrates schematically a comparison between the electrical conductivity of semiconductors having a forbidden energy gap of the order of 1 eV (e.g. silicon and germanium) and the conductivity of insulators having an energy gap of 5 eV. This comparison, based on a unique statistical thermodynamics approach, shows that :

1. The intrinsic carriers, produced by thermal excitation of electrons (or holes) in band-to-band transitions cannot contribute significantly to conduction in insulators, even at relatively high temperatures (500 K).

2. Extrinsic carriers produced by thermal excitation of electrons (or holes) from impurity levels do not contribute to conduction at or below room temperature. At higher temperatures (500 K), the conductivity due to thermal ionization of impurities may reach the limit of detection, of the order of 10^{-21} $(\Omega\text{-m})^{-1}$.

In practice, steady currents can be measured through samples of insulating materials subjected to a high electric field ($E \geqslant 10^7$ V m^{-1}). Typical values of the steady current flowing through 100 μm thick polymer samples of 10^{-3} m^2 surface area under an applied field of 10^7 V m^{-1} are of the order of 10^{-12} Å at room temperature. This corresponds to a conductivity σ of the order of 10^{-16} $(\Omega\text{-m})^{-1}$ which, although small, is much higher than that predicted in Table 11.

The steady-state current is usually non-ohmic, however, and increases faster than linearly with the applied field (typically as E^n with $2 \leqslant n \leqslant 3$ over limited range of applied fields). This raises the basic question of the physical significance of the conductivity, which will be discussed later.

TABLE 11 - Comparison of semiconductors and insulators

	Physical Property	Typical semiconductor e.g. Ge, Si, etc.	Typical insulator e.g. NaCl, Polymer, etc.	
	Optical absorption edge $\lambda(\mu m)$	1.5 (opaque)	$\leqslant 0.25$ (transparent)	
	Forbidden energy gap E_G (eV) $E_G = \frac{hc}{e\lambda} = \frac{1.24}{\lambda}$	0.8	$\gg 5$	
Intrinsic conduction both carriers assumed mobile		T = 300 K	T = 300K	T = 500K
	Free carrier density (m^{-3}) $np = N_o^2 \exp\left(\frac{-E_G}{kT}\right)$, $N_o = \left(\frac{2\pi mkT}{h^2}\right)^{3/2}$ $n = p = N_o \exp\left(\frac{-E_G}{2kT}\right)$	2.8×10^{18}	10^{-18}	1
	Free carrier mobility $\mu(m^2V^{-1}s^{-1})$	10^{-4} - 1	$\leqslant 10^{-8}$	
	Conductivity $\sigma = n\, e\, \mu\ (\Omega\text{-m})^{-1}$	4.5×10^{-5} - 0.45	$\leqslant 10^{-45}$	$\leqslant 10^{-27}$
	Effective mass ratio m^*/m_o	0.1	1	
Extrinsic (impurity) conduction ionized impurities assumed non-mobile	Optical permittivity $K = n^2$	16	2.5	
	Ionization energy from Bohr hydrogenic model $E_i = R(m^*/m_o)K^{-2}$ (eV)	5×10^{-3}	2	
	Impurity concentration $N(m^{-3})$	10^{18} - 10^{24}	$\leqslant 10^{26}$	
	Ionized impurity concentration $N_i = N \exp\left(-\frac{E_i}{kT}\right)$ (m^{-3})	10^{18} - 10^{24} Most centres are ionized at R.T.	$\leqslant 2.10^{-9}$	$\leqslant 10^{5}$
	Extrinsic conductitivy $(\Omega\text{-m})^{-1}$	1.6×10^{-5}-$1.6\ 10^{5}$	$\leqslant 10^{-35}$	$\leqslant 2 \times 10^{-22}$

The hyper-linear character of the current suggests that carriers of other origin than that discussed in Table 11 participate in the conduction.

Carriers of non-thermal origin are produced when high-energy particles such as cosmic rays collide with atoms and molecules in the samples, causing ionization.

The number of electronic charges produced in this way each second in unit volume (m^3) of a material of density ρ (kg m^{-3}) by a radiation field of intensity Φ (Rad s^{-1}) and energy W (eV) is

$$N = 6.24 \times 10^{17} \rho \Phi / W \qquad \text{(IX.1)}$$

The charge carriers generated by ionizing collisions contribute to a current which saturates with increasing field at a value corresponding to the transfer between the electrodes of all the radiation-excited carriers.

The integrated energy flux for cosmic radiation on the earth's surface is about 3×10^{-10} J m^{-2} s^{-1}, i.e. 400-500 particles per m^2 per second per sterad. This has been shown to produce a measurable conduction in highly-purified liquid alkanes and cryogens at moderately high fields ; however, other mechanisms are required to produce the observed superlinear conductance at very high fields. These are the charge injection processes (Schottky and Field-emission) which are discussed below.

IX.2. Injection processes

Among the field-controlled injection processes which may contribute to the observed currents flowing through insulators, Schottky and tunnel injections only will be discussed.

1. Schottky injection

This is the enhancement of thermionic emission arising from a lowering of the potential barrier which occurs in the combined applied and image fields.

Fig. 47 shows the potential energy diagram of the cathodic interface with an applied field E . The origin of the potential is the "vacuum level", ie. the potential in eV. of an electron, removed to an infinite distance from the cathode.

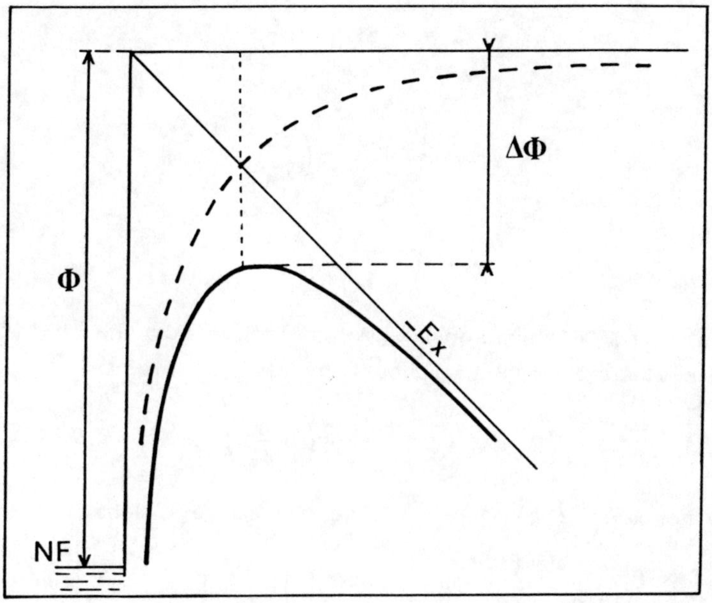

Fig. 47 - Energy diagram for Schottky injection.

The potential at a distance x from the cathode is the algebraic sum of:

1. the potential V_a due to the applied field

$$V_a = - (-1)E\,x \qquad \text{(IX.2)}$$

is represented by a straight line of negative slope, because the field E is negative. From now on, we consider only the magnitude of E and let $E = |E|$, so that $V_a = - Ex$.

2. the potential V_i due to the "image" of the electron in the metal. Since the distance between the electron and its image is $2x$, the attractive force is:

$$F_i = \frac{1}{4\pi\varepsilon}\,\frac{e^2}{4x^2}$$

so that the attractive potential is:

$$V_i = -\frac{1}{4\pi\varepsilon}\,\frac{e}{4x}$$

The resultant potential

$$V(x) = - Ex - \frac{e}{16\pi\varepsilon x} \qquad \text{(IX.3)}$$

has a maximum

$$V_m = - \Delta\Phi = - \left(\frac{eE}{4\pi\varepsilon}\right)^{1/2} \qquad \text{(IX.4)}$$

at a depth

$$x_m = \left(\frac{e}{16\pi\varepsilon E}\right)^{1/2} \qquad \text{(IX.5)}$$

Using the Richardson-Dushman relation for the thermionic current density emitted in vacuo without any applied field:

$$i_T = AT^2 \exp\left(- \frac{\Phi}{kT}\right) , \qquad \text{(IX.6)}$$

Φ now becomes $(\Phi - \Delta\Phi)$, so that the current density is:

$$i(E) = i_T \exp \frac{\Delta\Phi}{kT} = i_T \exp\left[\frac{1}{2kT}\left(\frac{eE}{\pi\varepsilon}\right)^{1/2}\right] \qquad \text{(IX.7)}$$

For a metal/vacuum interface with Φ of the order of 4 eV, i_T is exceedingly small at room temperature, since $\exp \frac{\Phi}{kT} = \exp(-160)$. However, the effective metal/dielectric work function can be much lower than 4 eV, as the electron extracted from the metal polarizes the dielectric and recovers from it the energy of polarization. If the "effective" work function is sufficiently low, (contacts have actually been made on crystals with an effective work function of 1 eV) the large exponential field-dependent factor in eqn. (IX.7) may compensate for the small first factor.

2. Tunnel injection

This process depends on the wave-like character of the electron. It is considered here for the well understood case of electron injection from a metal into vacuo at the absolute zero of temperature.

Figure 48 shows a potential energy diagram similar to that of the preceding figure. However, whereas the image potential was predominant earlier, it is now of minor importance, and can be neglected in a first approximation. Consequently, the potential is limited here to the straight line $V = -Ex$.

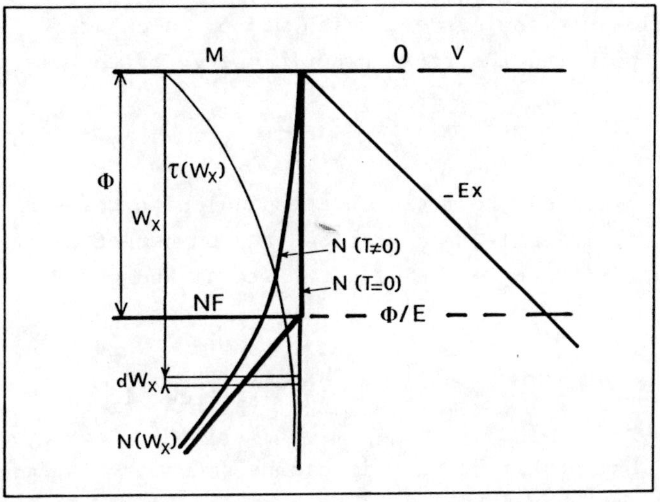

Fig.48 - Potential energy diagram for tunnel injection of electrons into vacuo.

Electrons in the metal colliding with the interface have momentum, but, according to Fermi-Dirac statistics, no electron has an energy higher than the Fermi energy Φ , and $p_x = \sqrt{2m\Phi}$ is the maximum value of the x-component of momentum.

According to classical mechanics, electrons cannot escape from the metal unless they acquire - from the lattice phonons for instance - sufficient energy to surmount the "potential barrier".

Wave mechanics predicts, however, that they can cross the barrier without acquiring extra energy, provided that the wavelength of the probability wave (De Broglie wave) associated with the electrons is larger than the thickness of the barrier.

For instance, at the Fermi Level, the thickness x_F of the potential barrier, given by $V(x) = \Phi$, is $x_F = \Phi/E$. Typically, for $\Phi = 5$ eV and $E = 5\times10^8$ V m^{-1}, $x_F = 10^{-8}$ m $= 100$ Å .

The De Broglie wavelength of an electron of momentum p is $\lambda = h/p$. Hence, if an electron hits the surface normally with a Fermi energy Φ , the associated wavelength is

$$\lambda = \frac{h}{\sqrt{2m\Phi}} = 6 \times 10^{-8}\text{m} = 600 \text{ Å}.$$

This wavelength being larger than the thickness of the potential barrier, the electron can "tunnel" through the barrier and contribute to a field-emission current which we now calculate.

We write the total kinetic energy of the electron in the form

$$W = \frac{Px^2}{2\ m} + \frac{Py^2 + Pz^2}{2\ m} = W_x + \frac{Py^2 + Pz^2}{2\ m} \tag{IX.8}$$

W_x corresponds to motion perpendicular to the surface.

From Fermi-Dirac statics, the number of electrons hitting 1 m^2 of the surface per unit time with a total energy W in the range $(W,W+dW)$ is

$$N(W)\ dW = \begin{cases} 0 & \text{if} \quad W \leqslant \Phi \\ \frac{4\pi mkT}{h^3}\ dW & \text{if} \quad W > \Phi \end{cases}$$

Using this result, it can be shown, by integration over p_y and p_z, that the number of electrons hitting 1 m^2 of the surface per unit time with a "normal energy" W_x in the range $(W_x, W_x + dW_x)$ is

$$N(W_x)\ dW_x = \begin{cases} 0 & \text{if} \quad W \leqslant \Phi \\ \frac{4\pi m}{h^3}\ (\Phi - W)dW_x & \text{if} \quad W > \Phi \end{cases}$$

$N(W_x)$ is known as the "supply function".

The total current density emitted by the metal into vacuo is the integral, from the vacuum level to the bottom of the valence band, of the product

$$dJ = e\ N(W_x)\ \mathcal{C}(W_x) dW_x \tag{IX.9}$$

where $\mathcal{C}(W_x)$, the "transmission" of the barrier for an electron of "normal energy" W_x, is the ratio of the intensity Ψ_T^2 of the transmitted wave to the intensity Ψ_i^2 of the incident wave.

With the triangular potential barrier considered here, the exact solution Ψ_T of the Schrödinger equation involves Airy functions which are cumbersome to handle.

In their classic treatment of 1928, Fowler and Nordheim used the Wentzel-Krämers-Brillouin (WKB) approximation, according to which the transmission of a potential barrier $V(x)$ for a level of energy W_x is:

$$\mathcal{C}(W_x) = \exp\left(- \frac{2\sqrt{m}}{h} \int_{x_1}^{x_2} \sqrt{V(x) - W_x}\ dx \right) \tag{IX.10}$$

where x_1 and x_2 are the solutions $V(x) - W_x = 0$. With the linear potential of the problem, the WKB approximation is:

$$\mathcal{C}(W_x) \simeq \exp\left(-\frac{8\pi\sqrt{2m}}{3\,e\,h\,E} W_x^{3/2}\right). \qquad \text{(IX.11)}$$

Assuming that the bottom of the valence band is at $(-\infty)$, and using the above approximation for $\mathcal{C}(W_x)$, J becomes:

$$J = \frac{4\pi m e}{h^3} \int_0^\infty (W_x - \Phi)\exp\left(-\frac{8\pi\sqrt{2m}}{3\,e\,h E} W_x^{3/2}\right) dW_x \qquad \text{(IX.12)}$$

Because of the factor $W_x^{3/2}$ in the exponential, the integral cannot be calculated analytically. However, the factor $W_x^{3/2}$ can be approximated, by Taylor expansion, to a polynomial of integer powers of W_x about any fixed value of W_x. Expanding about $W_x = \Phi$,

$$W_x^{3/2} = \Phi^{3/2} + \frac{2}{3}(W_x - \Phi)\,\Phi^{1/2} + \ldots \qquad \text{(IX.13)}$$

Introducing eqn.(IX.13) for W_x in J gives:

$$J = \frac{4\pi m e}{h^3} \exp\left(-\frac{8\pi\sqrt{2m}}{3\,e\,h\,E}\Phi^{3/2}\right) \int_\Phi^\infty (W_x - \Phi)\exp\left[-\frac{4\pi\sqrt{2m\Phi}}{e\,h\,E}(W_x - \Phi)\right] dW_x. \qquad \text{(IX.14)}$$

Now, the integral is easy to calculate, and the final result is:

$$J = \frac{e^3 E^2}{8\pi h \Phi} \exp\left(-\frac{8\pi\sqrt{2m}}{3\,e\,h\,E}\Phi^{3/2}\right) \qquad \text{(IX.15)}$$

If e, h and m are replaced by their numerical values in S.I. units,

$$J(\mathrm{Am}^{-2}) = 1.54\times10^{-10}\,\frac{E^2}{\Phi} \exp\left(-6.83\times10^9\,\frac{\Phi^{3/2}}{E}\right) \qquad \text{(IX.16)}$$

with Φ in eV and E in V m^{-1}.

This relation predicts that the emitted current density - negligibly small if E is lower than 10^8 V m^{-1} - increases sharply for higher fields. If $\frac{J}{V^2}$ is plotted against $\frac{1}{V}$, a straight line can actually be obtained over a decade in $\frac{1}{V}$, and the slope of the line gives the expected value of Φ.

It should be mentioned here that field-emission microscopy, which is based on the observation - on a fluorescent screen - of the emission pattern produced by a sharp metal tip at a high negative voltage with respect to the screen, constitutes both a brilliant confirmation of the above theory and a powerful metallurgical technique.

If the vacuum is replaced by a dielectric material, and, in particular, by a condensed one, the situation is much more complex. As mentioned earlier, the presence of the dielectric affects the work function and the shape of the potential barrier in a manner which is not yet fully understood.

X. SPACE-CHARGE-LIMITED, INJECTION-CONTROLLED CONDUCTION

We have seen previously that charges can be injected (or captured) by the electrodes under the influence of the field. These charges generate their own field, and a steady state is reached when the local field at the interface becomes equal to the field required for the injection (or capture) of the charges.

In this chapter, the steady-state, space-charge-limited current is considered, first in the case of a plane-parallel sample of infinite area (to avoid edge effects), then in configurations of practical importance. The following assumptions are made throughout:

1. The material has no intrinsic conductivity, but charges can be injected into it by one of the electrodes (usually electrons at the cathode) and captured by the other.

2. The injected charges have a uniform mobility μ which is independent of the local field.

3. The dielectric permittivity ε is unaffected by the presence of the injected charges.

X.1. Plane-parallel configuration. Mott's relation

According to the above assumptions, and to the equations of Section VII.9, the steady-state uniform current density across the sample is:

$$j = \mu \rho(x) E(x) + \mu \frac{kT}{e} \frac{d\rho(x)}{dx} \tag{X.1}$$

where $\rho(x)$ and $E(x)$ are respectively the charge density of the injected carriers and the field at a depth x. These quantities are functions of x, whereas j is uniform, by definition.

The second term of j is the diffusion current density previously considered in the theory of Space-charge relaxation (Section VII.9). In this term, Einstein's relation for the diffusion constant has been used.

The quantities $\rho(x)$ and $E(x)$ are again related, as seen in Section VII.9, by Poisson's equation:

$$\frac{dE(x)}{dx} = \frac{\rho(x)}{\varepsilon} \tag{X.2}$$

Combining eqns. (X.1) and (X.2) gives

$$j = \frac{\varepsilon\mu}{2}\left(\frac{dE^2}{dx} + \frac{2kT}{e}\frac{d^2E}{dx^2}\right) \tag{X.3}$$

This is a second-order differential equation, the solution of which involves two constants of integration, determined by two boundary conditions.

Straightforward integration of eqn.(X.3) gives

$$j(x+\lambda) = \frac{\varepsilon\mu}{2}\left(E^2 + \frac{2kT}{e}\frac{dE}{dx}\right) \tag{X.4}$$

where the length λ is the first constant of integration. However, two equations are required to determine the unknown quantities j and λ .

Whereas the integration of eqn.(X.4) involves Airy functions which are cumbersome to handle, it becomes elementary if the diffusion term can be neglected, as proposed by Mott. The validity of Mott's approximation is questionable, as discussed below, but the results are of such physical and historical importance that several electrode configurations are treated here using this approximation.

The usual argument to justify "a priori" the neglect of the diffusion term rests on the idea that the charge Q introduced in an open capacitor generates, by influence on the plates, a charge $Q' = \varepsilon V/L$ of magnitude comparable to Q . This is correct if Q is a thin charge layer in the vicinity of an electrode, but incorrect if Q is distributed uniformly across the capacitor, or symmetrically with respect to the midplane.

Using the approximation:

$$\frac{dE}{dx} \approx \frac{\Delta E}{\Delta x} ,$$

together with Poisson's equation and the above property of Q' ,

$$\frac{dE}{dx} \simeq \frac{E(L) - E(o)}{L} = \frac{Q}{\varepsilon V} \simeq \frac{V}{L^2} = \frac{E^2}{V} \tag{X.5}$$

so that eqn.(X.4) becomes

$$j(x+\lambda) \simeq \frac{\varepsilon\mu}{2} E^2\left(1 + \frac{2kT}{eV}\right) \tag{X.6}$$

At ambient temperature, $\frac{2kT}{e}$ is about 0.05 volts. Hence, the second term of the parenthesis is negligible compared to unity, provided that $V \gg 1$ volt. This is the case in practically any current measurement, except with very thin films of molecular thickness.

Consequently, provided that the problem raised by the "a priori" justification of Mott's approximation is not overlooked, this approximation can be used to reduce eqn. (X.4) to

$$j(x + \lambda) = \frac{\varepsilon\mu}{2} E^2 \quad . \tag{X.7}$$

One relation between j and λ is given by the boundary condition on the applied field:

$$\int_0^L E(x)\, dx = E_a L \tag{X.8}$$

where the magnitude of the applied field E_a is V/L.

Using eqn.(X.7) for E, the boundary condition given by eqn.(X.8) becomes:

$$E_a = \left(\frac{8jL}{9\mu\varepsilon}\right)^{1/2} \left[\left(1 + \frac{\lambda}{L}\right)^{3/2} - \left(\frac{\lambda}{L}\right)^{3/2}\right] \tag{X.9}$$

A special case of paramount interest is that for which the field at the injecting electrode - which is now assumed to be the cathode - is zero. This occurs when the negative space-charge injected into the material reduces the cathode field just sufficiently to cancel the applied field ($E(0) = 0$ and $\lambda = 0$).

Letting $\lambda = 0$ in eqn.(X.9) gives

$$E_a = \left(\frac{8jL}{9\varepsilon\mu}\right)^{1/2} \tag{X.10}$$

or

$$\boxed{j = j_s = \frac{9}{8}\,\varepsilon\,\mu\,\frac{E_a^2}{L}} \tag{X.11}$$

This simple equation is known as "Mott's relation". It gives the maximum, or saturation, current density (subscript s) which can flow through a dielectric sample of permittivity ε when electrons of mobility μ are injected from the cathode. The existence of a steady injected current j_s with $E(0) = 0$ implies that electrons from the cathode can leak into the material in the absence of a cathode field, regardless of the potential barrier. In spite of the paradox raised here, the subsequent development will show that Mott's relation applies very well if $E(0)$ is small.

In fact, this equation has been surprisingly well verified in various materials, mostly in wide-gap semiconductors, but also in organic crystals of high intrinsic resistivity, such as anthracene fitted with properly injecting contacts. For such materials, Mott's relation provides an estimation of the mobility of the injected carriers. This mobility is usually low (in the range 10^{-12}-10^{-6} $m^2\ V^{-1}\ s^{-1}$) because the carriers spend most of their time in shallow traps, so that the measured value is, in fact, an effective mobility.

The voltage, field and charge-density distributions across the sample given by Mott's approximation are shown in Fig.49 , for various degrees of injection. In the absence of injection (a) , the potential is linear in x , the field is uniform, and the charge density is uniformly zero in the sample. If electrons are injected by the cathode (b) , the potential is distorted downwards, the field distribution is parabolic, and the charge-density is negative throughout the sample, with a maximum value at the cathode. If the current density approaches its saturation value (c) , the peak of the parabolic field distribution approaches the cathode and the charge density there increases to infinity.

Strictly speaking, this limiting case is physically irrelevant, since the diffusion current cannot be neglected with respect to the conduction current wherever the local field is small, as is the case at the cathode if $E(o) = 0$. However, as the field gradient at the cathode is now infinite, the field remains small only over a very small depth.

In the same way, the total charge stored by unit area of sample

$$\int_o^L \rho(x)\, dx = \varepsilon E(L) = \frac{3}{2}\, \varepsilon E_a \qquad (X.12)$$

remains finite, in spite of the fact that the charge density $\rho(x)$, from Fig.49.III , diverges for $x = 0$.

A better understanding of this topic and of the problems raised by the cathode field can be gained by the use of reduced variables. Let $\mathcal{E}$ be the ratio between the cathode field and the average applied field $\mathcal{E} = E(o)/E_a$, and i the ratio $i = j/j_s$ between the actual current density and the saturation-current density j_s given by Mott's relation

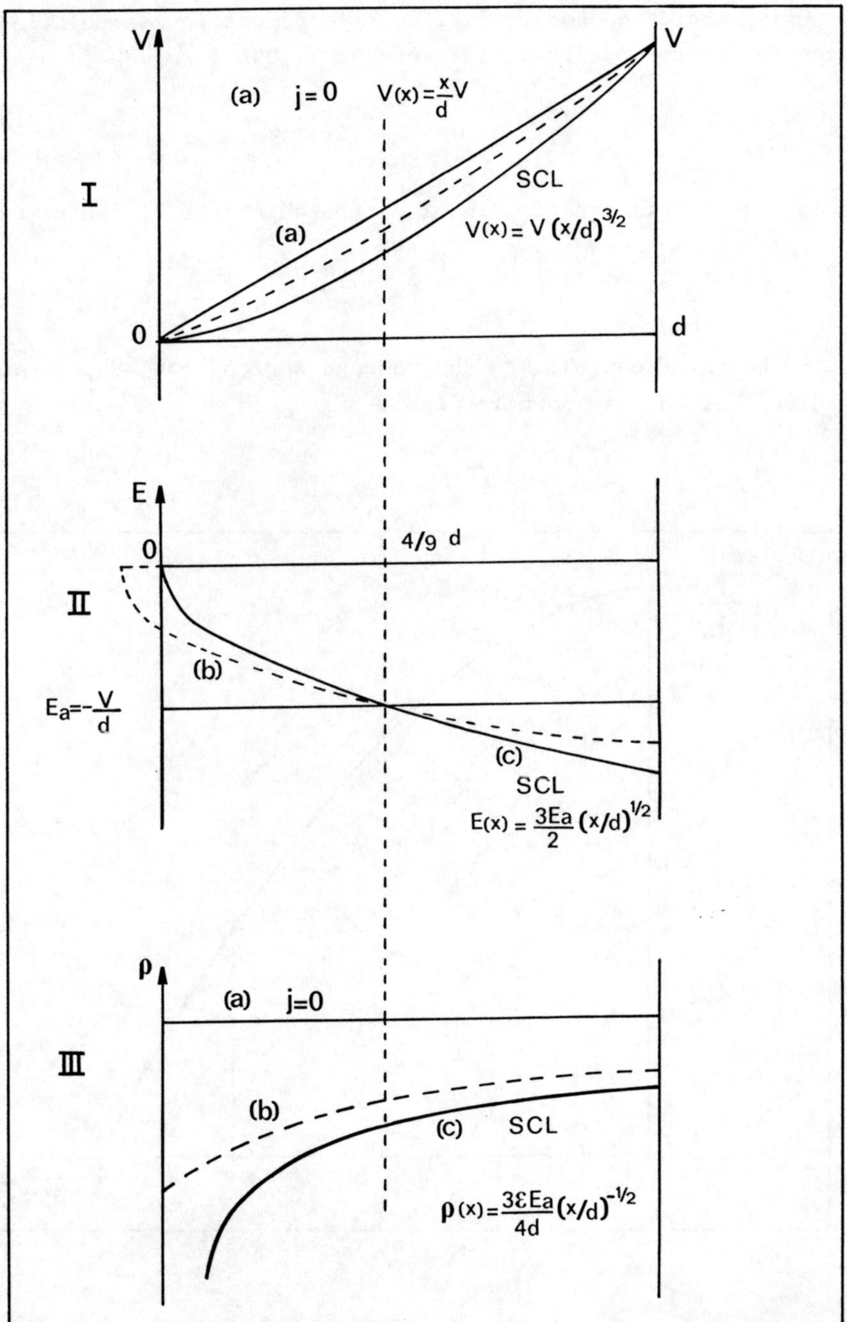

Fig.49 - Representation of V(x), E(x) and ρ(x) in the case of unipolar injection from the cathode on the left. The dashed line is the intermediate case between no injection (j=0) and space-charge limiting (SCL).

Using these reduced variables, some algebraic manipulations - suggested to the reader as an exercise - transform eqn.(X.9) into

$$i = \left(i + \frac{4}{9}\,\mathcal{E}^2\right)^{3/2} - \left(\frac{4}{9}\mathcal{E}^2\right)^{3/2} \qquad (X.13)$$

This is a quadratic equation in i , from which $i(\mathcal{E})$ can be extracted:

$$i(\mathcal{E}) = \frac{1}{2} - \frac{2}{3}\mathcal{E}^2 + \frac{1}{2}\left[1 - \frac{8}{3}\mathcal{E}^2 + \frac{64}{27}\mathcal{E}^3 - \frac{16}{27}\mathcal{E}^4\right]^{1/2} \qquad (X.14)$$

By means of eqn.(X.14), the reduced current can be plotted versus the reduced field, as shown in Fig.50.

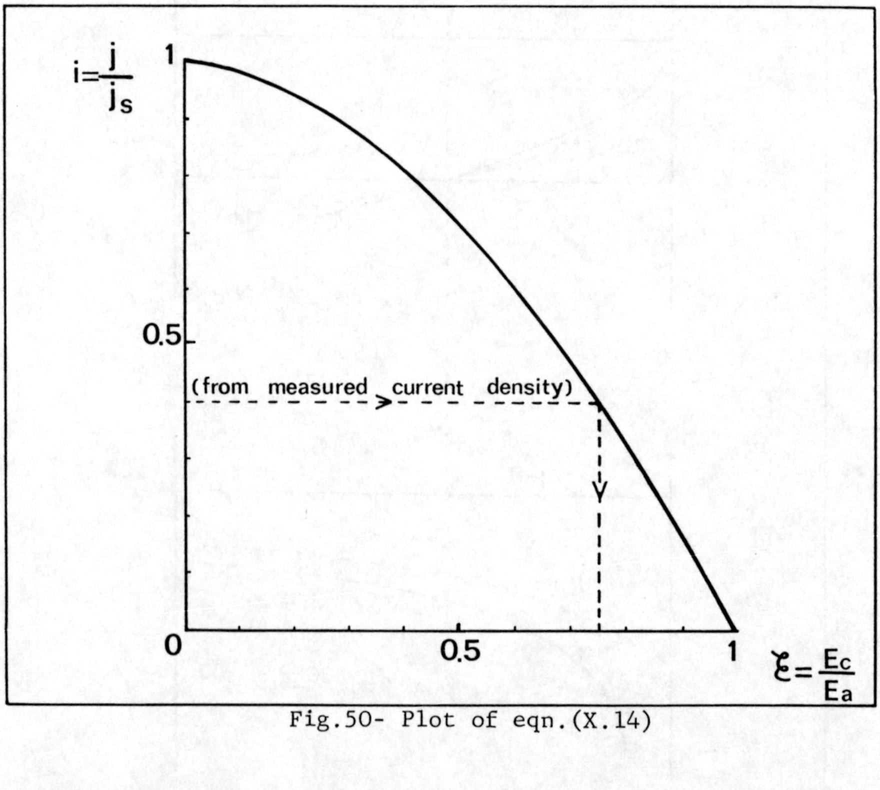

Fig.50- Plot of eqn.(X.14)

As expected, the reduced field is unity when no current is injected, and the reduced current is unity when the reduced field is zero (saturation). The horizontal slope of the curve at the saturation point is quite significant. It shows that, even if the current is not fully saturated, its value is close to that of the space-charge-limited current density j_s . In other words, Mott's relation is a fair approximation to the current density, even if the cathode field is not exactly zero, and this remark stresses the importance of this relation.

Furthermore, when unipolar injection takes place in a material of known ε and μ , the value of i can be obtained by dividing the measured current density by that given by Mott's relation. Figure 50 can be used to find the cathode field and hence to obtain some idea of the injection process.

Problem

The transfer times.

1. Calculate the transfer time (ie. the time required by a carrier to move from the cathode to the anode) in the absence of space-charge (negligible injection). Assume that the carriers have no initial velocity at the cathode. Call this transfer time t_o .
2. Calculate the transfer time t_s in the case of space-charge saturation ($\lambda = 0$). Show that the transfer time t in the intermediate case of non-saturating injection is between t_o and t_s . Plot t/t_o versus the reduced current density $i = j/j_s$.

X.2. Cylindrical configuration

This configuration is very common in practice: it is that of any high voltage cable, for instance. The same treatment as was used for the plane geometry can be carried out, replacing the uniform current density $j(x)$ by the uniform radial current density:

$$I = 2\pi r\, j(r) = 2\pi r \mu \rho(r)\, E(r)$$

expressed in Amps. per metre rather than in Amps. per square metre. Poisson's equation in this geometry is:

$$\frac{dE(r)}{dr} + \frac{E}{r} = \frac{\rho(r)}{\varepsilon} \qquad (X.15)$$

and elimination of $\rho(r)$ between eqns.(X.14) and (X.15) gives

$$\frac{dE^2(r)}{dr} + \frac{2\,E^2(r)}{r} = \frac{2a}{r} \qquad (X.16)$$

where $$a = \frac{I}{2\pi\varepsilon\mu} \tag{X.17}$$

Integration of eqn.(X.16) gives:

$$E(r) = \left(a + \frac{C}{r^2}\right)^{1/2} \tag{X.18}$$

where the constants a and C are related by the boundary condition

$$\int_{r_i}^{r_o} E(r)\,dr = V \tag{X.19}$$

C can be positive or negative, and the special case $C = 0$ corresponds to a uniform field.

Fig.51 is the exact analogy, in the cylindrical configuration, of Fig.49.II which showed $E(x)$ for the plane-parallel configuration, with the assumption that the carriers (electrons) are injected at the surface of the inner cylinder of radius r_i.

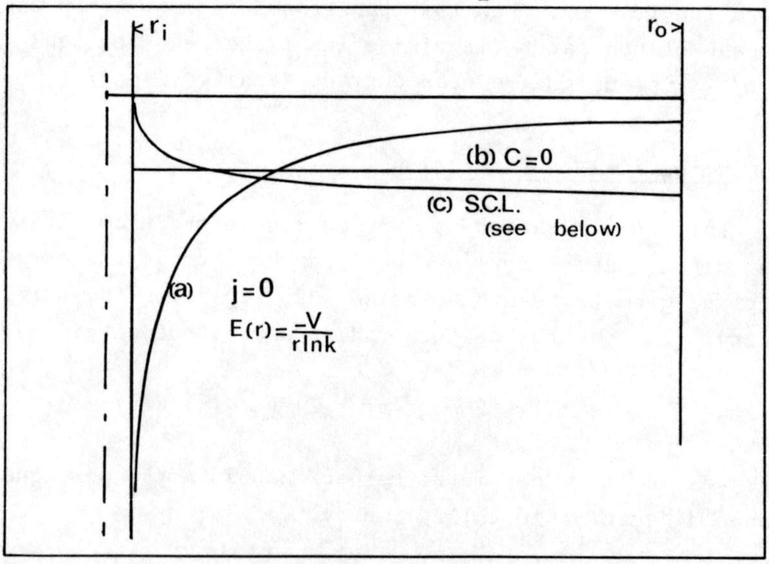

Fig.51- Plot of the field in the coaxial configuration.
a) j = 0 , b) C = 0, c) S.C.L.

Problem

In the particular case (C) of space-charge limitation (SCL) and assuming further that $k = r_o/r_i \gg 1$, derive eqns.(X.20) to (X.22):

$$j_s = \frac{2\pi\varepsilon\mu}{r_i^2}\left(\frac{V}{k-\frac{\pi}{2}}\right)^2 \quad (\text{Amp m}^{-1}) \qquad (X.20)$$

$$E_s(r) = -\left(1-\frac{r_i^2}{r^2}\right)^{1/2} \frac{V}{r_i(k-\frac{\pi}{2})} \qquad (X.21)$$

$$\rho_s(r) = -\frac{2V}{rr_i(k-\frac{\pi}{2})}\left(1-\frac{r_i^2}{r^2}\right)^{-1/2} \qquad (X.22)$$

X.3. Spherical configuration

The total current I is now related to the current density $j(r) = \mu\rho(r)E(r)$ by the equation:

$$I = \Omega\, r^2\, j(r) \qquad (X.23)$$

where Ω is the solid angle of injection.

For a complete sphere, $\Omega = 4\pi$, but this condition cannot be realized in practice, since the current collected from the outer sphere is carried to the inner sphere by a conductor which cannot make contact with the outer sphere. Therefore, we shall not assign any particular value to the solid angle Ω, and write:

$$I = \Omega\,\mu\, r^2\, \rho(r)E(r)\ . \qquad (X.24)$$

Poisson's equation for the spherical configuration is

$$\frac{1}{r^2}\,\frac{d(r^2E)}{dr} = \frac{\rho(r)}{\varepsilon} \qquad (X.25)$$

Elimination of $\rho(r)$ between eqns.(X.24) and (X.25) gives

$$E(r)\,\frac{d(r^2E)}{dr} = \frac{I}{\Omega\mu\varepsilon} \qquad (X.26)$$

This can be written as a linear first order equation in $E^2(\frac{1}{r})$:

$$\frac{dE^2}{d(1/r)} - 4\,\frac{E^2}{(1/r)} + \frac{2I}{\Omega\varepsilon\mu} = 0 \quad (5) \tag{X.27}$$

Eqn.(5) can be integrated, giving:

$$E^2 = \frac{2I}{3\Omega\varepsilon\mu}\left(\frac{1}{r}\right) + K^2\left(\frac{1}{r}\right)^4 \tag{X.28}$$

where the constant of integration K is such that

$$\int_{r_i}^{r_o} E(r)\,dr = V$$

Unfortunately, the integral

$$I = \int (a + br^3)^{1/2}\,\frac{dr}{r}$$

has no analytic solution, so that exact equations for the current-voltage characteristics cannot be obtained, and approximations have to be used. From such approximations, it can be shown that the current-voltage characteristic is quadratic over a wide range of applied voltage.

X.4. Point-Plane configuration

The difficulties met with in relatively simple configurations (Cf. previous sections) leave us with little hope of solving the problem of space-charge controlled injection in more complex electrode configurations.

Nevertheless, the general equations will be written, and discussed for the point-plane configuration, which is commonly used in high-field conduction and breakdown experiments, since the divergent field favours charge injection from the tip.

Using Poisson's equation in its general form:

$$\vec{\nabla}.\vec{E} = \rho/\varepsilon \tag{X.29}$$

the current density becomes

$$\vec{j} = \mu\varepsilon(\vec{\nabla}.\vec{E})\,\vec{E} \tag{X.30}$$

Applying to $\vec{j}$ the steady-state condition of charge conservation gives:

$$\vec{\nabla}.\left\{ (\vec{\nabla}.\vec{E})\ \vec{E} \right\} = 0 \qquad (X.31)$$

or

$$(\vec{\nabla}.\vec{E})^2 + \vec{E}.\ \vec{\nabla}\ (\vec{\nabla}.\vec{E}\ ) = 0 \qquad (X.32)$$

A tip of axis z can be reasonably well regarded as a hyperboloid of revolution described by the equations:

$$\begin{aligned} r &= a \cos\zeta \sinh \eta \\ z &= a \sin\zeta \cosh \eta \end{aligned} \qquad (X.33)$$

These equations represent two orthogonal sets of confocal hyperboloids (η variable and ζ fixed) and ellipsoids (η fixed and ζ variable). Two particular values of ζ ($\zeta = 0$ and $\zeta = \zeta_o \leqslant \frac{\pi}{2}$) define the equipotentials ; $\zeta = 0$ is the plane electrode (z = 0), at zero potential for instance, and $\zeta = \zeta_o$ defines the tip, raised to a high voltage V (Fig.52).

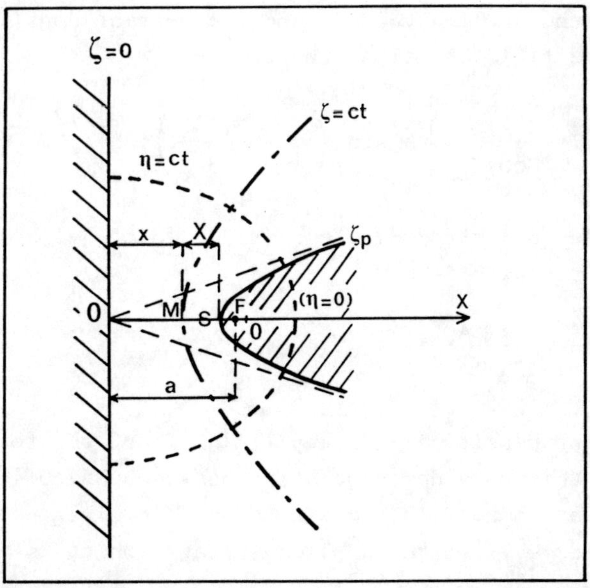

Fig.52 - Hyperbolic approximation of the point/plane configuration.

In the absence of injection from the tip, the equipotentials are hyperboloids of parameters ζ $(0 \leqslant \zeta \leqslant \zeta_0)$ (cf. Durand - General bibliography). In the presence of injection, the equipotential surfaces and the field lines are slightly perturbed, but the tip axis remains the line of maximum field, by symmetry. Along this axis, the field is a function of ζ only, and eqn.(X.32) can be written in the form

$$\frac{d^2E^2}{d\,\zeta^2} - 5\tan\zeta\,\frac{dE^2}{d\zeta} - 4\,E^2 = 0 \tag{X.34}$$

Letting $Y = E^2\cos^4\zeta$, eqn.(X.35) becomes:

$$\frac{d^2Y}{d\zeta^2} + 3\tan\zeta\,\frac{dY}{d\zeta} = 0 \tag{X.35}$$

Equation (X.35) gives, after straightforward integration:

$$Y = A\sin\zeta\left(1 - \frac{\sin^2\zeta}{3}\right) + B \tag{X.36}$$

A and B, the constants of integration, are determined as before, by the boundary conditions on the applied voltage and the current density.

Coming back to the field E along the tip axis,

$$E(\zeta) = \frac{1}{\cos^2\zeta}\left[A\sin\zeta\left(1 - \frac{\sin^2\zeta}{3}\right) + B\right]^{1/2} \tag{X.37}$$

Now, as can be seen by introducing eqn.(X.37) into eqn.(X.34), the current density on the tip axis is

$$j(\zeta) = \frac{\mu\,\varepsilon\,A}{2a\cos^2\zeta} \tag{X.38}$$

The emitted current density at the tip $(\zeta=\zeta_0)$ is of the form $j(\zeta_0) = f\left[E(\zeta_0)\right]$, and this provides a relation between the constants A and B. The second relation between these constants is given by the boundary condition $\int E dz = V$ on the applied voltage, which is here:

$$\int_0^{\sin\zeta_0} E(\zeta)\cos\zeta\,d\zeta = V/a. \tag{X.39}$$

Here, we meet the same type of problem as before, since the

integral cannot be calculated analytically. However, adequate approximations show that over a range of voltages the current again varies as the square of the applied voltage.

XI. FIELD-INDUCED INTRINSIC CONDUCTION

In addition to the charge carriers injected from the electrodes, carriers may also be produced in the bulk of the materials under the action of the electric field on the constituent atoms and molecules.

Among others, the Poole-Frenkel effect and the field-induced dissociation processes will now be discussed.

XI.1. The Poole-Frenkel effect

This mechanism is quite similar in nature to the Schottky injection treated in Section IX.2. However, whereas Schottky injection resulted from the field-induced lowering of the potential barrier at the metal/dielectric interface, the Poole-Frenkel effect, due to the combined potentials of the applied field and the ionization centre, implies a decreased ionization energy and hence an increased concentration of ionized carriers in the presence of the field.

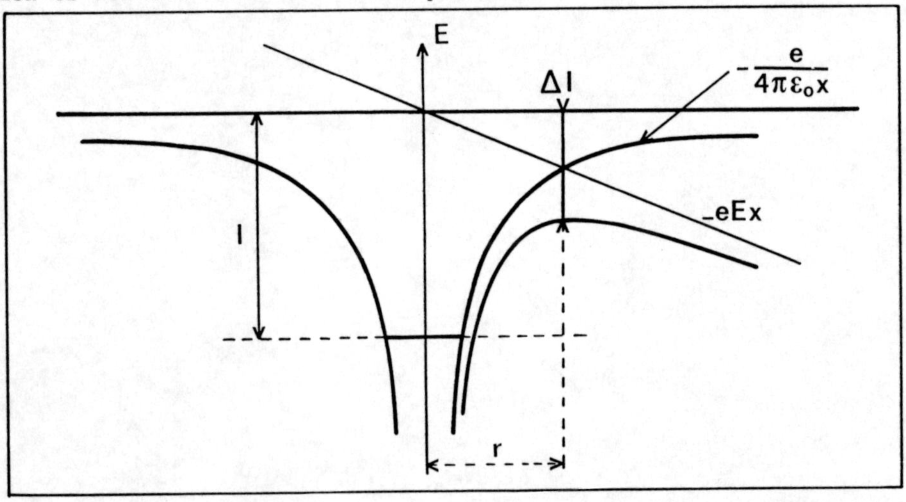

Fig.53 - Schematic energy diagram for the Poole-Frenkel effect.

Fig.53 represents the potential energies in the plane defined by the ionized centre and the applied field. It shows that, in the applied field E, the ionization potential I becomes $(I - \Delta I)$. For the same value of E, the lowering ΔI:

$$\Delta I = \left(\frac{eE}{\pi\varepsilon}\right)^{1/2} \qquad (XI.1)$$

is twice the barrier lowering $\Delta\Phi$ for Schottky injection, because the image potential in 1/4x is now replaced by a Coulomb potential in 1/r.

The current density corresponding to the Poole-Frenkel effect depends upon the nature of the carriers.

In solid phases, the ionized carriers are usually not mobile. Hence, Poole-Frenkel conduction in these materials is due exclusively to electrons from the ionizable centres, of concentration

$$n(E) = n(0) \exp \frac{\Delta I}{kT} = n(0) \exp \left[\frac{1}{kT} \left(\frac{eE}{\pi\varepsilon} \right)^{1/2} \right] \qquad (XI.2)$$

The graph representing $\log_{10}$ of the conductivity versus $\sqrt{E}$ is a straight line of slope $\frac{0.434}{kT}\sqrt{\frac{e}{\pi\varepsilon}}$.

In liquid phases, both carriers may contribute equally to the conduction and, in this case, the mass-action law predicts that the concentration of each type of carrier is:

$$n_+(E) = n_-(E) \propto \left[\exp \frac{1}{kT} \left(\frac{eE}{\pi\varepsilon} \right)^{1/2} \right]^{1/2}$$

or

$$n_+(E) = n_-(E) \propto \exp \frac{1}{2kT} \left(\frac{eE}{\pi\varepsilon} \right)^{1/2} \qquad (XI.3)$$

The exponential factor of eqn.(XI.3) is exactly the same as that of eqn.(IX.7) for Schottky injection, and this makes any clear-cut distinction between Schottky injection and the Poole-Frenkel effect extremely difficult in practice.

XI.2. Field-induced dissociation

The dissociation of neutral molecules in the presence of an electric field:

$$AB \underset{}{\overset{K(E)}{\rightleftharpoons}} A^+ + B^-$$

can also be considered as one of the possible sources of charge carriers. In fact, extending this model to the case of very weak electrolytes subjected to a very high field, Onsager has shown that the field-dependent dissociation constant K(E) is given by a formula which differs from eqn.(XI.2) only in a pre-exponential factor containing $E^{-3/4}$. This factor varies slowly with the field, as compared to the exponential term.

It is quite fascinating to realize that such profoundly different models, with only a thermally activated process as a common feature, give similar results, although this does not help to clarify our understanding of field-induced conduction in insulators.

XI.3. General formulation of conduction with generation and recombination of carriers

This very general problem was considered in 1903 by J.J. Thomson, in his pioneering investigation of conduction through gases. In principle, it can be applied to any insulator or semiconductor. It cannot be solved analytically, but powerful computers can be used to find numerical solutions in practically any situation.

We consider a material between plane-parallel electrodes of infinite dimensions, in which carriers of charges $\pm e$ can be generated by a thermal and/or field-assisted process, producing one carrier of each sign. We assume that the carriers may recombine by collisions.

Let $n_+(x)$ be the steady concentration of positive charge carriers of mobility μ_+, $n_-(x)$ the concentration of negative charge carriers of mobility μ_-, and $q_\pm = e\, n_\pm$ the magnitudes of the corresponding charge densities.

The applied voltage is assumed large with respect to 0.1 Volt, so that diffusion currents may be neglected (cf. Section X.1).

We are looking for the steady field distribution $E(x)$ and the variation of the total current with the applied voltage.

The steady-state current density is uniform:

$$i = i_+(x) + i_-(x). \tag{XI.4}$$

From now on, the depth x will be omitted for simplicity in the equations. With this simplified notation, eqn.(XI.4) becomes:

$$i = (\mu_+ q_+ + \mu_- q_-)E \tag{XI.5}$$

and Poisson's equation relating $q_\pm$ to the local field is

$$\varepsilon \frac{dE}{dx} = q_+ - q_- \tag{XI.6}$$

From eqns.(XI.5) and (XI.6), q_+ and q_- can be extracted:

$$q_\pm = \frac{1}{\mu_+ + \mu_-}\left(\frac{i}{E} \pm \varepsilon\mu_\mp \frac{dE}{dx}\right) \tag{XI.7}$$

In the steady-state, with uniform total current density, conservation of the positive and negative charges gives:

$$\frac{di_+}{dx} = -\frac{di_-}{dx} = \gamma - \alpha q_+ q_- \qquad \text{(XI.8)}$$

In eqn.(XI.8), γ is the rate of generation of the positive and negative carriers. As mentioned above, it can, in principle, depend on the field and therefore on x, but in the following calculation, γ is assumed uniform. The recombination coefficient α, expressed in m^3 per Coulomb per second, is also assumed uniform.

Using the expressions for i_+ and i_-, eqn.(XI.8) becomes:

$$\mu_+ \frac{d}{dx}(q_+E) = -\mu_- \frac{d}{dx}(q_-E) = \gamma - \alpha q_+ q_- \qquad \text{(XI.9)}$$

Multiplying both sides of Poisson's equation by E gives:

$$\frac{\varepsilon}{2}\frac{dE^2}{dx} = (q_+ - q_-)E \qquad \text{(XI.10)}$$

Differentiation of eqn.(XI.10)gives:

$$\frac{d^2E^2}{dx^2} = \frac{2}{\varepsilon\bar{\mu}}(\gamma - \alpha q_+q_-) \qquad \text{(XI.11)}$$

where

$$\frac{1}{\bar{\mu}} = \frac{1}{\mu_+} + \frac{1}{\mu_-}$$

Finally, using in eqn.(XI.11) the values of q_- and q_- given by eqn.(XI.7), we obtain:

$$\frac{d^2E^2}{dx^2} = \frac{2\gamma}{\varepsilon\bar{\mu}} - \frac{2\alpha}{\varepsilon\bar{\mu}(\mu_+ + \mu_-)^2}\left(\frac{i}{E} + \varepsilon\mu_-\frac{dE}{dx}\right)\left(\frac{i}{E} - \varepsilon\mu_+\frac{dE}{dx}\right) \qquad \text{(XI.12)}$$

Simple algebraic manipulations transform this equation into:

$$\frac{d^2E^2}{dx^2} = \frac{2\gamma}{\varepsilon\bar{\mu}} + \frac{2\alpha\varepsilon}{\mu_+ + \mu_-}\left(\frac{dE}{dx}\right)^2 - \frac{\alpha i}{(\mu_+ + \mu_-)^2E^2}\left(\frac{2i}{\varepsilon\bar{\mu}} + \frac{\Delta\mu}{\bar{\mu}}\frac{dE^2}{dx}\right) \qquad \text{(XI.13)}$$

with $\Delta\mu = \mu_- - \mu_+$.

With the dimensionless variables

$$y = \left(\frac{E}{E_a}\right)^2 , \quad \text{where } E_a = V/L \text{ is the applied field}$$

and $v = x/L$,
eqn.(XI.13) takes the form:

$$yy'' - Ay'^2 + By' - Cy + D = 0 \qquad (XI.14)$$

In this equation, the dimensionless factors are:

$$A = \frac{\alpha \varepsilon}{2(\mu_+ + \mu_-)} , \quad B = \frac{\alpha iL}{(\mu_+ + \mu_-)^2 E_a^2} \frac{\Delta\mu}{\bar{\mu}} , \quad C = \frac{2\gamma L^2}{\varepsilon \bar{\mu} E_a^2} , \quad D = \frac{2\alpha i^2 L^2}{\varepsilon \bar{\mu}(\mu_+ + \mu_-)^2 E_a^4}$$

Eqn.(XI.14) is rather cumbersome, and a search for its solution can only be attempted in very special cases. Some of these are considered below.

1. Only one of the carriers is mobile

Suppose, for instance, $\mu_+ \rightarrow 0$, then $\bar{\mu} \rightarrow 0$. The terms containing B, C and D in eqn.(XI.14) dominate over the others, and the equation reduces to a first-order one:

$$y' - c\,y + d = 0 \qquad (XI.15)$$

with $c = \frac{2\gamma L \mu}{\alpha i \varepsilon}$ and $d = \frac{2\,i\,L}{\varepsilon\,\mu\,E_a^2}$

where μ is the mobility of the negative carriers.

The general solution of eqn.(XI.15) can be written in the form:

$$y = a^2(1 + \lambda\, e^{cv}) \qquad (XI.16)$$

with

$$a^2 = \frac{\alpha i^2}{\gamma \mu^2 E_a^2}$$

The integration constant λ is determined by the boundary condition:

$$\int_0^L E dx = E_a L \qquad (XI.17)$$

which, written in terms of y and v, is:

$$\int_0^1 \sqrt{y}\, dv = 1 \qquad (XI.18)$$

If the expression for y given by eqn.(XI.16) is used in eqn.(XI.18), a relationship is obtained between a, c and λ of the form:

$$f(i, E_a, \lambda) = 0 \qquad \text{(XI.19)}$$

In the present case where y is given by eqn.(XI.16), the integral in eqn.(XI.18) can be solved analytically. This is left as an exercise for the reader.

From eqn.(XI.19) it is possible, in principle, to extract $\lambda = \lambda(i, E_a)$ and, by introducing this expression for λ into eqn.(XI.16), the field distribution is obtained:

$$E = E(x, i, E_a)$$

Finally, the boundary condition at the cathode

$$i = i\left[E(o)\right] = i[E(o,i, E_a)] \qquad \text{(XI.20)}$$

provides the current-voltage characteristics $i = i(E_a)$.

2. Both carriers have the same mobility

If $\mu_+ = \mu_- = \mu$, then $\bar{\mu} = \frac{\mu}{2}$ and $\Delta\mu = 0$. Hence, $B = 0$ and eqn. (XI.14) becomes:

$$yy'' - Ay'^2 - Cy + D = 0 \qquad \text{(XI.21)}$$

with

$$A = \frac{\alpha E}{4\mu}, \qquad C = \frac{4\gamma L^2}{\varepsilon\mu E_a^2} \quad \text{and} \quad D = \frac{\alpha i^2 L^2}{\varepsilon\mu^3 E_a^4}$$

By letting $y'^2 = Y$, eqn.(XI.21) can be transformed into a linear equation in $Y(y)$:

$$\frac{dY}{dy} - 2A\frac{Y}{y} + \frac{2D}{y} - 2C = 0 \qquad \text{(XI.22)}$$

Eqn.(XI.22) can be integrated by parts, and its general solution, for $A \neq 1/2$ is:

$$Y = y'^2 = K\,y^{2A} + \frac{2C}{1-2A}\,y + \frac{D}{A} \qquad \text{(XI.23)}$$

where K is the first integration constant.

The second integration:

$$v = \int_{y(o)}^{y} \left(K\, y^{2A} + \frac{2C}{1-2A}\, y + \frac{D}{A} \right)^{-1/2} dy$$

together with the boundary condition eqn.(XI.15) can only be carried out analytically for $A = 0$ or $A=1$, so that a numerical solution is usually required.

Problem

The case $A = \frac{1}{2}$.

1. Show that eqn.(XI.19) with $A = \frac{1}{2}$ takes the form:

$$\frac{d}{dy}\left(\frac{Y}{y}\right) + \frac{2D}{y^2} - \frac{2C}{y} = 0$$

and that integration gives:

$$Y = y'^2 = K\, y + 2\, C\, y \ln y + 2\, D$$

where K is the first integration constant.

2. This equation cannot be further integrated, because of the logarithm However if $E(x)$ is not very different from the applied field $(E(x) \sim E_a)$, y is close to unity and the logarithmic term vanishes. An approximate equation may then be written:

$$y' = (K\, y + 2\, D)^{1/2}$$

solve for $y(v)$ in this case, and discuss the conditions under which y remains close to unity.

If the rate of generation of carriers does not depend on the field, the current cannot exceed the value which corresponds to the transfer, to the electrodes, of all the generated carriers.

Consequently, the fact that the measured current never saturates with increasing applied voltage implies a field-dependent generation of carriers.

XII. DIELECTRIC STRENGTH

In a sufficiently high electric field, materials break down, either abruptly or gradually, with the formation of one or several conducting spots in the sample. This process is irreversible in a solid. The breakdown field depends, of course, on the nature and composition of the sample, but also on its shape, on its environment, and on the way in which the field is applied (d.c., a.c., ramp voltage, square pulses, etc...).

For condensed insulators, the breakdown fields usually observed range from below 10^5 V cm^{-1} to about 5 x 10^6 V cm^{-1}. Although apparently large on a macroscopic scale, these fields are in fact very low on the atomic scale. 10^6 V cm^{-1} represents 10^{-2} V $\mathring{A}^{-1}$, and this clearly indicates that - except under very special laboratory conditions, breakdown <u>never</u> results from the direct action of the field on the atoms and molecules of the material. As an exercise, the reader might calculate by how much a field of 10^6 V cm^{-1} displaces the proton from the centre of a hydrogen atom according to the spherical model studied in Section III.2. He should find that the displacement is less than 0.1% of the atomic radius.

Thus, electrical breakdown is a collective phenomenon, whereby energy is communicated to the constituent molecules by other particles such as electrons or ions which have acquired enough energy in the field.

In this chapter, we shall first describe the most extrinsic type of breakdown, the thermal runaway, then more intrinsic types, such as the intrinsic and the avalanche-breakdown processes, which might be labelled under the same heading of collision breakdown.

Although it is convenient both for the writer and for the reader to classify related phenomena, so that each one is ascribed a section of its own, the distinction may be somewhat arbitrary and possibly misleading. The case of the breakdown processes is a striking example of this. Regardless of the initial step, the development of the breakdown event involves a local dissipation of energy from the excited carriers into heat, through excitation of lattice vibrational modes. Only the mechanism initiating the breakdown process differs significantly in the various cases, but as soon as this initiating step has occurred, breakdown develops in a very similar way, and observation of the breakdown spot does not help much in deciding what type of breakdown has taken place.

Recording of the current at the beginning of the breakdown event constitutes a potential means of identifying the process involved, but raises enormous experimental difficulties.

XII.1. Thermal breakdown

1. Definition

This occurs whenever the Joule heating in the sample cannot be extracted fast enough by conduction and/or convection, so that the sample temperature rises until permanent damage occurs.

The heat capacity of rather thick samples may be such that breakdown occurs several hours after the application of the voltage.

In the elementary treatment which follows, two simplifying assumptions are made, namely:

a) The sample is always isothermal (This is fairly well justified in the case of thin samples).

b) The heat flux carried out of the sample at temperature T to the ambient medium at T_o is of the form $\Gamma(T-T_o)$, Γ being a heat transfer coefficient accounting for both conduction and convection processes.

A steady temperature can be reached and breakdown does not therefore occur if the product $\Gamma(T-T_o)$ balances the power IV lost by unit area of the sample. I is the magnitude of the uniform current density, given by:

$$I = \sigma(T)\,\frac{V}{L} \qquad \text{(XII.1)}$$

where $\sigma(T)$ is the temperature-dependent conductivity of the sample, V the applied voltage and L the sample thickness. The equilibrium equation is:

$$VI = \Gamma(T-T_o) \qquad \text{(XII.2)}$$

To simplify the calculations, we assume here that the conductivity $\sigma(T)$ is independent of the applied field. This is not the case, particularly at high fields, as explained in Chapters X and XI. Furthermore, although the conductivity is usually a thermally-activated process of the form:

$$\sigma(T) = \sigma_\infty \exp\left(-\frac{U}{kT}\right), \qquad \text{(XII.3)}$$

we shall use the simpler form:

$$\sigma(T) = \sigma_o \exp \lambda(T-T_o) \qquad \text{(XII.4)}$$

which is a good approximation to eqn.(XII.3) if T is not too different from T_o. By comparing eqns.(XII.3) and (XII.4) it is found that, under these conditions, $\lambda \simeq U/kT_o^2$.

From eqns.(XII.1), (XII.2) and (XII.4), the steady-state equation may be written:

$$\frac{V^2}{L} \sigma_o \exp \lambda(T-T_o) - \Gamma(T-T_o) = 0 \qquad \text{(XII.5)}$$

This transcendental equation can be solved by plotting its two terms separately on the same graph, as a function of $(T-T_o)$. This is done in Fig.54. For the case $V=V_1$, the exponential curve crosses the straight line at two points A and B. If the initial temperature is T_o, the sample heats up to T_A. If, accidentally, the sample temperature rises above T_A, the rate of heat loss exceeds the rate of generation, and the sample cools down to T_A. Consequently, T_A is a stable temperature. Provided that T_A is lower than the temperature at which the sample degrades, the sample can withstand the voltage V_1 indefinitely, and breakdown does not occur.

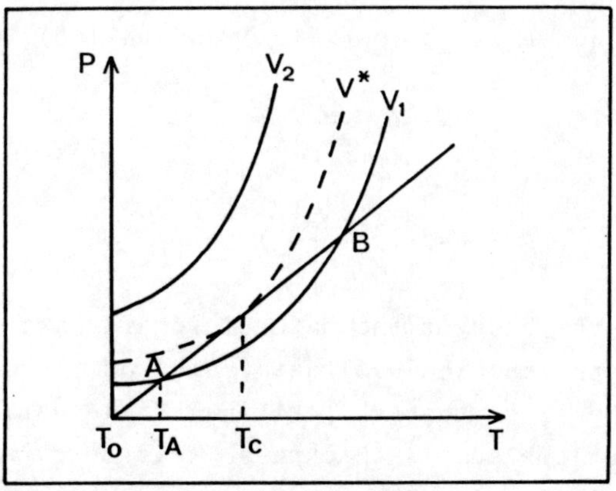

Fig.54 - Power-temperature graph for thermal breakdown. The straight line is the heat carried away from the sample, and the curves represent the heat dissipated in the sample for various values of the applied voltage.

Conversely, for $V = V_2$ (= 2V, for instance), the exponential curve lies above the straight line. There is no intercept, so that the sample temperature keeps increasing until breakdown occurs, when the degradation temperature is reached. The time between application of the voltage and breakdown can be calculated from the dynamic heat equation:

$$C_v \frac{dT}{dt} = VI - \Gamma(T-T_o) \tag{XII.6}$$

where C_v is the heat capacitance of the sample (at constant volume). This delay, involving C_v and T can be quite long, as stated earlier.

It is obvious that the critical voltage V_c for which breakdown eventually occurs at an infinite time is simply that for which the straight line is a <u>tangent</u> to the exponential curve, and the critical temperature T_c is that at which the curve contacts the straight line, provided that no deterioration occurs before this temperature is reached.

The critical voltage, current density and temperature result from the condition that the point of slope λ on the exponential curve most lie on the straight line, and our assumptions on the conductivity simplify the calculation, which gives:

$$T_c = T_o + \frac{1}{\lambda} \tag{XII.7}$$

Using eqns.(XII.7) in (XII.2) and (XII.5), we find:

$$V_c = \left(\frac{L\Gamma}{e\,\sigma_o\lambda} \right)^{1/2} \tag{XII.8}$$

and

$$I_c = \left(\frac{e\,\sigma_o\Gamma}{\lambda L} \right)^{1/2} \tag{XII.9}$$

where e is the base of the natural logarithms. From eqn.(XII.8), it appears that the critical voltage varies as the square root of the sample thickness, or that the critical field varies as the inverse square root of the sample thickness. This thickness dependence of the breakdown field is often observed, and may be considered, together with the relatively long associated time-lag, as a criterion of thermal breakdown. A good example of this type of breakdown is obtained with Langmuir films such as barium stearates, margarates, palmitates, etc. Fig.55 shows the thickness dependence of barium palmitate as measured by Agarwal and Srivastava. From this plot, the breakdown field varies as $L^{-0.54}$, over more than one decade of thickness. This is in good agreement with the power (-0.5) predicted by the theory of thermal breakdown.

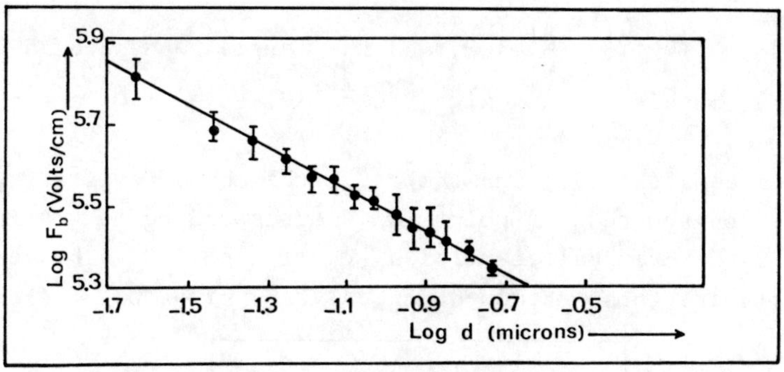

Fig.55 - Log-log plot of breakdown field F_b vs. dielectric thickness d for barium palmitate (230-1850 Å).
From V.K. Agarwal and V.K. Srivastava Thin Solid Films, 13 (1972) S23-S24.

2. <u>I-V Characteristics and thermal instability</u>

It is instructive to calculate the I-V steady-state characteristics of the sample. This can be done by eliminating the term $(T-T_o)$ between equation (XII.1) or its equivalent:

$$I = \frac{V\sigma_o}{L} \exp \lambda (T-T_o) \tag{XII.1b}$$

and the power equilibrium equation (XII.2).

The result of this elimination is:

$$I = \frac{V\sigma_o}{L} \exp \frac{\lambda VI}{\Gamma} \tag{XII.10}$$

or

$$VI = \frac{\Gamma}{\lambda} \ln \frac{IL}{\sigma_o V} \tag{XII.11}$$

Although these implicit relations cannot be solved analytically for V or I, differentiation of both sides of eqn.(XII.11) and rearrangement gives the local slope of the characteristics:

$$\frac{dI}{dV} = \frac{I}{V} \frac{\Gamma + \lambda IV}{\Gamma - \lambda IV} \tag{XII.12}$$

From eqn.(XII.12), it is obvious that, if V is the imposed variable, instability occurs $\left(\frac{dI}{dV} = \infty\right)$ for

$$IV = \frac{\Gamma}{\lambda} = (IV)_c \tag{XII.13}$$

If $V < V_c$, $IV < \frac{\Gamma}{\lambda}$, $\frac{dI}{dV} > 0$, and the sample behaves as a variable resistance, but if V_c is high, then $IV > \frac{\Gamma}{\lambda}$, $\frac{dI}{dV} < 0$ and the sample breaks down.

From eqns.(XII.11) and (XII.12), the characteristics given in Fig. 56 can be constructed. At point C, corresponding to the instability ($IV = \frac{\Gamma}{\lambda}$), the critical quantities V_c, I_c and T_c can be calculated, and the results are those obtained directly and given by eqns.(XII.7) to (XII.9).

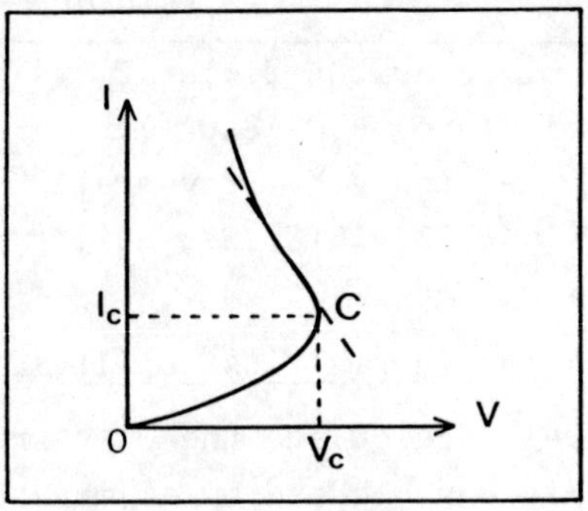

Fig.56 - IV characteristics including the negative resistance branch.

If the measurement is made with a voltage generator of relatively low output impedance, point C corresponds to thermal breakdown, and the branch with negative slope on the characteristics cannot be observed. On the other hand, with a current generator - for instance a high-voltage generator with a series resistance - the negative branch of the characteristics can be observed, and the slope at the inflexion, for $VI = \sqrt{3}\Gamma/\lambda$, sets a lower limit of 0.66 R_o for the series resistance, where R_o is the resistance of the sample at room temperature.

This technique is quite useful in the developing field of switching devices of the "ovonic" type, and the newcomer in the field should be aware of the advantages of current generators over the more usual voltage generators.

As a conclusion to this section, it should be remembered that the critical conditions for thermal breakdown (V_c and T_c) sharply depend on the heat transfer coefficient Γ, and consequently on the nature and morphology of the electrodes. The more readily they carry out the heat dissipated in the sample, the higher is the thermal-breakdown field Therefore, even though relatively high breakdown fields are observed in the case of thin films, the thermal-breakdown strength is not really an intrinsic property of the material itself.

3. Case of thick films with a non-uniform temperature

The assumption of an isothermal sample, which was justified in the case of thin films, does not hold for thick films, in which the heat generated in the bulk is less easily dissipated than the heat generated close to the electrodes. As a result, the temperature is non-uniform, as can be seen in Fig.57. $T_1 = T(\pm L/2)$ is the temperature at the interface, $T_m = T(0)$ the maximum temperature in the middle plane, and T_o the ambient temperature. Breakdown can occur if T_m reaches a value for which the material deteriorates, and this criterion might be more severe than the previous one.

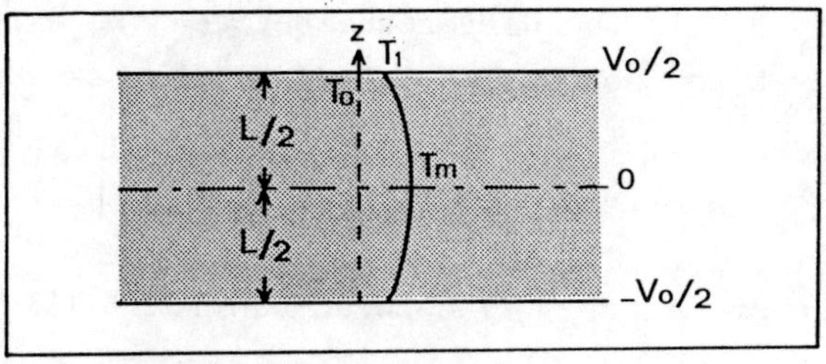

Fig.57 - The temperature profile in a thick sample.

In the steady-state, the continuity equations for the current density and the heat flux respectively are:

$$I = \sigma(T)\ E\ (z) = \sigma\ (T)\ \frac{\partial V}{\partial z} \tag{XII.14}$$

$$\frac{\partial}{\partial z}\ (K\ \frac{\partial T}{\partial z}\ ) = -\ \sigma\ (T)\ E^2\ (z) = -\ I\ \frac{\partial V}{\partial z} \tag{XII.15}$$

In these equations, $E\ (z)$ is the field at z, and K is the thermal conductivity of the material. Integration of eqn.(XII.15) gives:

$$K\ \frac{\partial T}{\partial z} + IV = 0 \tag{XII.16}$$

The integration constant at the right-hand side of eqn.(XII.16) is zero because of the symmetry of the configuration. I can be eliminated between eqns.(XII.14) and (XII.16):

$$\frac{2K}{\sigma(T)}\ \frac{\partial T}{\partial z} = -\ \frac{\partial V^2}{\partial z} \tag{XII.17}$$

and this equation can be integrated between the mid-plane z=0 ($T=T_m$) and the plane at temperature T:

$$V^2 = 2K \int_T^{T_m} \frac{dT}{\sigma(T)} \tag{XII.18}$$

If eqn.(XII.4) for $\sigma(T)$ is used, the integral is easily calculated, and the result is:

$$V^2 = \frac{2K}{\sigma_1 \lambda} \left[\exp - \lambda(T-T_1) - \exp(-\lambda \Delta T) \right] \tag{XII.19}$$

where $\Delta T = T_m - T_1$ is the difference between the extreme temperatures in the sample. At the boundaries, eqn.(XII.19) takes the form:

$$V_o^2 = \frac{2K}{\sigma_1 \lambda} \left[1 - \exp(-\lambda \Delta T) \right] \tag{XII.20}$$

and elimination of $\exp(-\lambda \Delta T)$ between eqns.(XII.19) and (XII.20) yields:

$$V^2(T) = \frac{V_o^2}{4} - \frac{2K}{\sigma_1 \lambda} \left[1 - \exp - \lambda(T-T_1) \right] \tag{XII.21}$$

From eqn.(XII.20), ΔT is an increasing function of the applied voltage, and breakdown may occur if ΔT exceeds some critical value.

If the sample is cooled efficiently, $T_1 = T_o$; otherwise, T_1 can be related to T_o by a steady-state power balance equation similar to eqn.(XII.2):

$$\int_0^{d/2} I\,E\,dz = I\,\frac{V_o}{2} = \Gamma'(T_1 - T_o) \tag{XII.22}$$

The coefficient Γ' is now the heat-transfer coefficient between the sample and the surroundings.

Ohm's law gives $V_o = 2RI$, where $2R$ is the sample resistance:

$$2R = 2\int_0^{L/2} \frac{dz}{\sigma} \tag{XII.23}$$

Finally, eqn.(XII.16) may be used to convert the integration over z into one over T :

$$V_o = 2\,K \int_{T_1}^{T_m} \frac{dT}{V(T)\,\sigma(T)} \tag{XII.24}$$

and this solves the problem, since the functions $\sigma(T)$ and $V(T)$ are given explicitly by eqns.(XII.4) and (XII.21) respectively.

In this treatment, T was assumed to be a function of z only. In practice, the samples are never perfectly homogeneous, so that the temperature is, in fact, a function of x and y as well as of z. These inhomogeneities may generate conducting areas, which build up perpendicular to the electrodes. Heat generation increases in these areas until local destruction occurs.

This "filamentary" breakdown is commonly observed in thin films, and may contribute to the "switching", from a high resistance to a low resistance state, of the "ovonics" devices made of thin amorphous layers of oxides or polymers.

XII.2. Intrinsic breakdown processes

1. Electron-phonon interaction. According to the famous Bloch theorem, electron waves do not interact with a perfect periodic lattice. In other words, in a perfect crystal, at the absolute zero of temperature, electrons excited into the conduction band (by photons, for instance, as there are no thermal electrons at absolute zero) should move freely in the lattice and consequently their mean free path should be infinite, provided that their own field does not distort the lattice. According to this statement, the breakdown field of a perfect crystal at 0 Kelvin should be vanishingly small. The fact that it is not so does not invalidate Bloch's theorem, but simply proves that electrons are always scattered, at vanishingly low temperatures, by lattice defects or by the distortion which they produce in the lattice.

This model is supported by the experimental results on the breakdown strength of alkali halide crystals, obtained mostly in Von Hippel's and Fröhlich's laboratories. These results clearly establish that the breakdown strength of these crystals indeed increases as the sample temperature is raised from 4 K to the ambient, before decreasing again at higher temperatures.

Assuming that the electrons are scattered exclusively by the phonons, the mean free path should vary as $T^{-3/2}$. Consequently, if breakdown occurs when the energy W reaches a critical value, the breakdown field $E = W/e\lambda$ should vary as $T^{3/2}$.

The scattering of electrons by thermal modes of vibration seems to be the only model which predicts an increase of the breakdown strength with increasing temperature.

The actual variation of the breakdown strength with temperature is not nearly as steep as predicted by phonon scattering, presumably because other scattering mechanisms are also operating, but the validity of the model is established. For instance, it is illustrated here by the results of Kaseta on the temperature dependence of the breakdown strength of pure potassium chloride, and of the same crystal in which 25% of the K^+ ions are replaced by Rb^+ ions (Fig.58). In both cases, the breakdown strength increases with increasing temperature.

Moreover, substitution of Rb^+ ions to K^+ ions increases the breakdown strength by about 20%. This is consistent with the fact that this substitution increases the lattice disorder. However, the slope of the curve E (T) should be smaller for the Rb^+-doped crystal, since impurity scattering is practically independent of temperature. This does not seem to be observed.

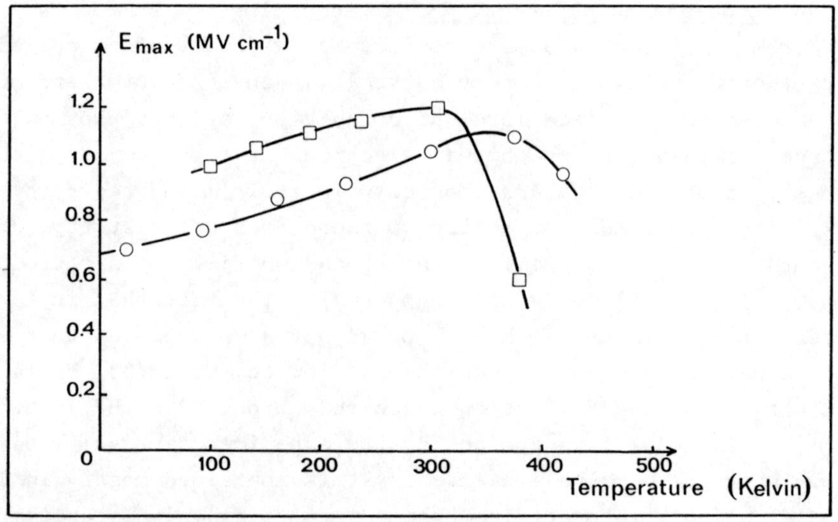

Fig.58 - Breakdown field versus absolute temperature.
o - pure KCl □ - 75% KCl - 25% RbCl.
After F. Kaseta and H.T. Li J.A.P. 37 2744 (1966).

The intrinsic breakdown model sketched above predicts a breakdown strength which is independent of the shape of the sample, and, in particular, of its thickness.

In fact, all the data show that the measured dielectric strength always decreases with increasing sample thickness. For instance, Fig.59 shows a synthesis of data on pure NaCl crystals at room temperature, obtained by Vorob'ev et al. (U.S.S.R) and Watson et al. (U.S.A). Whereas the Russian workers investigated very thin samples (3×10^{-4} to

2×10^{-3} cm), the Americans covered the thicker range (3×10^{-3} to 2×10^{-1} cm).

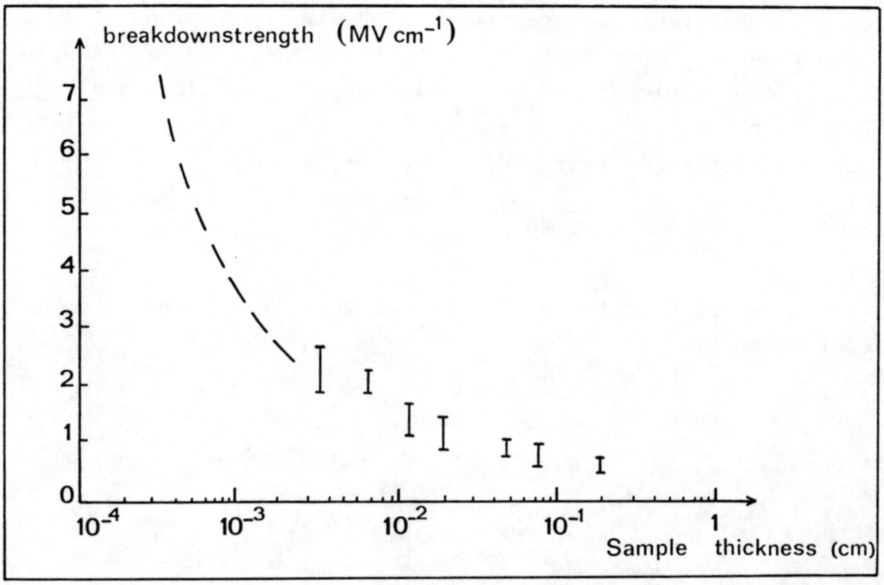

Fig.59 - Thickness dependence of breakdown strength of NaCl.
----- Vorob'ev et al., Radiotekhn. i Elektro. 7 1523 (1962)
I Watson et al., I.E.E.E. Electr. Insul. 1 30 (1965)

The agreement between these results is remarkable, and shows a strong thickness dependence of the breakdown strength, of the form:

$$E^* = \frac{E_o}{\ln(L/L_o)} \qquad \text{(XII.25)}$$

where E_o and L_o are constant field and thickness respectively. This does not agree with the $L^{-0.5}$ dependence predicted by the thermal breakdown model, and consequently another model has to be sought.

2. The avalanche breakdown model . The first coherent model for the electrical discharge in gases was developed by Townsend in 1901 . If an electron injected at the cathode can gradually acquire, from the field, energy equal to the ionization energy I of the molecules, the first collision it then undergoes is ionizing, and this event releases a second electron.

During this collision, the original electron loses most of its energy, so that the net result is two electrons of low energy. In principle each of these electrons can now gain energy in the field and produce, by collision, two low-energy electrons and so on.

Assuming that the free time of all the electrons in the avalanche is exactly the same, a total of about 2^n electrons are released when the original electron has undergone n ionizing collisions, (of course, this number is only approximate, due to the dispersion of the free time).

If we again assume that only ionizing collisions limit the free path of the electrons, this free path is $\lambda_I = I/eE$, and the number of electrons released over a distance $n\lambda_I = nI/eE$ is:

$$N = 2^n = \exp(n\ln 2) \qquad \text{(XII.26)}$$

According to the definition of the first Townsend multiplication coefficient α,

$$N = \exp(\alpha n\lambda_I) \qquad \text{(XII.27)}$$

From eqns.(XII.26) and (XII.27), it follows that:

$$\alpha = \frac{\ln 2}{\lambda_I} = \ln 2 \frac{eE}{I} \qquad \text{(XII.28)}$$

According to Townsend's model, breakdown in gases cannot occur unless "secondary" electrons are released at the cathode as a consequence of the primary ionization process, and breakdown occurs if

$$\gamma\left[\exp(\alpha L) - 1\right] \geqslant 1$$

γ is the "second" Townsend coefficient for electrons released at the cathode by the impact of positive ions and/or photons created by the primary event.

Using eqn.(XII.28) for α , the above breakdown criterion becomes:

$$\gamma\left[\exp\left(\ln 2 \frac{eV}{I}\right) - 1\right] \geqslant 1 \qquad \text{(XII.29)}$$

Equation (XII.29) is a voltage criterion, and consequently, it is not correct.

The model implies that the only inelastic collisions where energy is exchanged between the incoming electron and the target molecules are the ionizing collisions. This is not true, except perhaps in gases in the vicinity of the Paschen minimum. It can be intuitively understood, and rigorously demonstrated with time-dependent quantum mechanics, that a minimum duration of the interaction between the electron and the target molecule is required for adequate efficiency of the interaction.

In other words, the incoming electron should not be too fast and hence not too energetic.

It is found that the highest efficiency of the interaction - or the largest capture cross-section of the target molecule - occurs when the incoming electron can excite vibrational modes in the molecules. This implies energies of the order of 0.1 eV, two orders of magnitude smaller than the ionization energy I, and means that most of the electrons actually undergo inelastic collisions with the molecules before they have acquired the ionization energy I from the field. In other words, the mean free path of the electrons is one or two orders of magnitude smaller than $\lambda_I = I/eE$.

3. Avalanche in condensed phases. The avalanche breakdown model for gases has been extended to the condensed phase.

In gases, the avalanche can develop enormously without producing current runaway, so that breakdown requires a secondary process (the process mentioned above).

In condensed phases, positive ions created by ionizing collisions in the bulk are either immobile or very slow. They can increase the cathode field, and thereby increase the avalanche rate until it eventually becomes destructive. They do not release secondary electrons, as in gases, by collision with the cathode.

On the other hand, the optical absorption coefficient of condensed materials in the energetic UV range is usually so high that secondary photoelectrons cannot be produced at the cathode. Consequently, no secondary cathode process, in the usual sense, can take place. However, the sidewise spreading of the primary electrons is considerably reduced, so that the electron - and hence current - density can reach values which destroy the material, by local thermal breakdown generated in the avalanche.

According to Seitz, the critical density in the avalanche is reached after 40 generations of primary ionization events. With our previous assumption (no inelastic collision before the ionizing event), the distance over which this occurs is :

$$40\lambda_I = 40 \frac{I}{eE} ,$$

and the breakdown criterion becomes $L \geqslant 40\lambda_I$. This again defines a breakdown voltage $V = 40I/e \simeq 400$ Volt, which is obviously much too small for a thick sample of condensed insulator.

This confirms that most electrons undergo inelastic collisions with the lattice (exchange of energy between the electron and the vibrational modes of the lattice) before acquiring the ionization energy, and this will now be discussed.

Let $P(t_I)$ be the probability for an electron to undergo no inelastic collision during the time t_I that it requires to acquire an energy I in the field E. From elementary classical dynamics, it can be shown that $t_I = \sqrt{2mI}/eE$.

With $I = 10$ eV and $E = 10^8$ V m^{-1}, t_I becomes 10^{-13}s, longer than the actual mean free time of electrons between inelastic collisions with the lattice, as deduced from the plasma resonance frequency.

For a given electron, the average time interval $\langle dt/dn_I \rangle$ between two subsequent ionizing collisions is the product of the actual mean free time τ between collisions, and the ratio of non-ionizing to ionizing collisions. This ratio is just the inverse of the probability $P(t_I)$ defined above. Consequently, the average number of ions produced per unit time by an electron is:

$$\left\langle \frac{d\, n_I}{dt} \right\rangle = \frac{P(t_I)}{\tau} \qquad \text{(XII.30)}$$

The multiplication coefficient α, which is the average number of ionizations produced per unit length by an electron, can now be written as:

$$\alpha = \left\langle \frac{d\, n_I}{d\lambda} \right\rangle = \frac{\left\langle \frac{d\, n_I}{d\, t} \right\rangle}{\left\langle \frac{d\lambda}{d\, t} \right\rangle} = \frac{P\,(t_I)}{\tau \bar{v}} \qquad \text{(XII.31)}$$

where $\bar{v}$ is a simplified notation for the average velocity $\left\langle \frac{d\lambda}{d\,t} \right\rangle$.

At the breakdown field, $\bar{v} = \mu E^*$, so that eqn.(XII.31) becomes:

$$\alpha = \frac{P\,(t_I)}{\tau \mu E^*} \qquad \text{(XII.32)}$$

According to Seitz, a sufficient condition for breakdown is that 40 ionizing events occur over the sample thickness L. Thus:

$$\alpha = 40/L \qquad \text{(XII.33)}$$

Combining eqns.(XII.32) and (XII.33) gives the breakdown field as a function of the probability $P(t_I)$:

$$E^* = \frac{L}{\tau\mu} \frac{P(t_I)}{40} \tag{XII.34}$$

4. Vibrational barrier

Both intrinsic and avalanche breakdown processes rely on the fact that a carrier of charge e moving freely over a distance λ in a field E acquires from the field an energy $W = eE\lambda$.

For example, if $\lambda = 1000$ Å , and $E = 10^6$ V cm^{-1}, W = 10 eV, which is the same order of magnitude as the ionisation energy. Hence, if a free carrier has not lost its energy by inelastic collision with lattice before it has travelled about 1000 Å, the collision which will end its free path is likely to be ionizing.

Von Hippel and Fröhlich have given different criteria for the onset of breakdown in crystal lattices, but a common feature of their models is the concept of "vibrational barrier", which extends to an electron-lattice configuration the interaction developed earlier between an electron and the modes of vibration of a target molecule. The interaction between the wave associated to the electron and the vibration waves (phonons) has its peak efficiency when the electron energy is a few tenths of an eV. In other words, the phonon cross-section for an electron accelerated in a lattice by the applied field is a barrier-like function of the energy, and the shape of this barrier determines the breakdown strength.

5. Calculation of $P(t_I)$

To evaluate $P(t_I)$, we can express $P(t + dt)$ in terms of P(t) in two different ways, and compare the results.

On the one hand, Taylor expansion of $P(t + dt)$ gives:

$$P(t + dt) \simeq P(t) + \left(\frac{dP}{dt}\right) dt \tag{XII.35}$$

On the other hand, we can write:

$$\begin{pmatrix}\text{Probability for}\\ \text{no collisions}\\ \text{during } (0 - t + dt)\end{pmatrix} = \begin{pmatrix}\text{Probability for}\\ \text{no collisions}\\ \text{during } (0 - t)\end{pmatrix} \times \begin{pmatrix}\text{Probability for}\\ \text{no collisions}\\ \text{during } (t - t + dt)\end{pmatrix}$$

The second factor above is just the complement of the probability $\frac{dt}{\tau}$ that a collision does occur in the interval dt, and simultaneous solution gives:

$$P(t + dt) = P(t) + \left(\frac{dP}{dt} \right) dt = P(t) \left(1 - \frac{dt}{\tau}\right)$$

This reduces to:

$$\frac{dP}{dt} = - \frac{P(t)}{\tau} \tag{XII.36}$$

and integration gives:

$$P(t_I) = \exp\left[-\int_0^{t_I} \frac{dt}{\tau} \right] \tag{XII.37}$$

The mean free time τ is a function of the velocity, since the capture cross-section of the vibration modes depends on the energy of the electron. Therefore, rather than t, it is better to use a variable v, related to t by the dynamical equation:

$$m dv = eE^* dt$$

With this new variable, $P(t_I)$ becomes:

$$P(t_I) = \exp\left[- \frac{m}{eE^*} \int_0^{v_I} \frac{dv}{\tau(v)} \right] \tag{XII.38}$$

Strictly speaking, the lower limit of the integral should be the thermal velocity, but equating this to zero does not appreciably affect the results.

The term $\frac{m}{e} \int_0^{v_I} \frac{dv}{d\tau(v)}$ in the integral is a field $\mathcal{E}$, which can be calculated if the electron scattering process - and hence $\tau(v)$ - is known. Using this field $\mathcal{E}$, eqn.(XII.38) becomes:

$$P(t_I) = \exp(-\mathcal{E}/E^*) \tag{XII.39}$$

Finally, combining eqns.(XII.34) and (XII.39) gives:

$$40\,\tau\mu E^*/L = \exp(-\mathcal{E}/E^*)$$

or

$$E^* = \mathcal{E} \left(\ln \frac{L}{40\,\tau\mu E^*} \right)^{-1} \tag{XII.40}$$

This implicit expression for E^* agrees fairly well with the results of Fig.59.

It has been shown recently that 38 generations, instead of 40, may be sufficient to produce breakdown ; however, the critical number of generations enters in a logarithmic term, and therefore the result is not very sensitive to its exact value.

XII.3. Effect of pulse duration

For several reasons which will be discussed briefly, the breakdown strength usually decreases with increasing duration of the field pulse. This is illustrated by the data of Fig.60, showing the breakdown field of Pyrex glass sheets as a function of temperature, for field pulses of durations varying between 10^{-5} and 30 s.

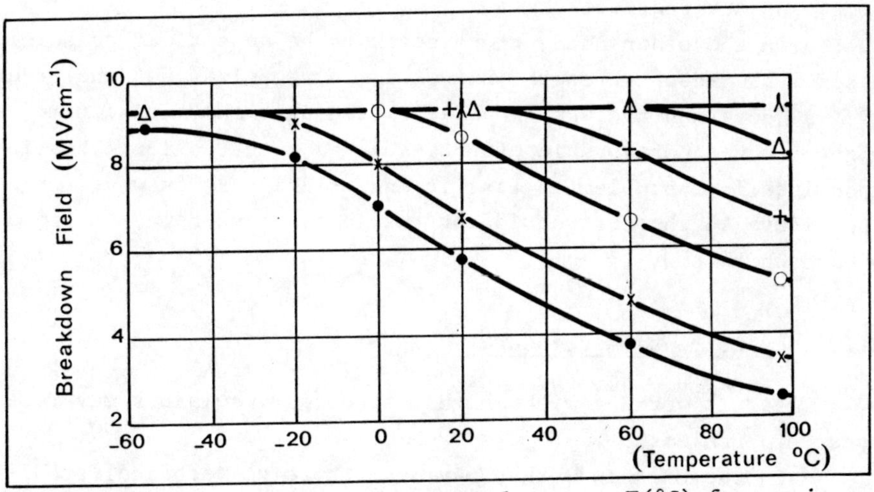

Fig.60 - Breakdown strength of Pyrex glass vs. T(°C) for various pulse widths : λ 10^{-5} s, Δ 10^{-4} s, + 10^{3} s, o 10^{-2} s. x 1 s, • 30 s , after J. Vermeer, Physica 20 313 (1954)

The pulse width dependence of the breakdown strength can be ascribed to various causes. Increasing the pulse width has three major consequences, namely;

(a) an increased generation of heat by the Joule effect in the sample, and consequently an increased probability of thermal breakdown at low fields.

(b) an increased space-charge due to injection and/or ion accumulation, and consequently an increased maximum over average field ratio.

(c) an increased swelling under the electrostatic pressure εE^2 (for $\varepsilon = 3\varepsilon_0$ and $E^* = 10^8$ V m^{-1}, this pressure is 2.6 bar). This increases the actual-over-calculated field ratio, since the sample may be thinner when it breaks down than in the absence of applied field.

This swelling is more frequent in plastics such as P.T.F.E. (Teflon) at high temperatures, and it contributes to a purely electrostatic transient current following the application of voltage ; but it has also been reported with harder materials, such as sodium chloride crystals.

These three mechanisms tend to lower the breakdown strength when longer pulses are applied. They may operate separately or together, combining their effects.

On the other hand, there seems to be an optimum pulse width for maximum breakdown strength. Extremely short pulses (of the order of 10^{-9} s. duration or less) have an exceedingly short rise time, and this might produce unfavorable effects such as electrocaloric heating, as described in a problem of Part 1, and/or sharp shock waves which may contribute to the destruction of the sample, and hence to premature breakdown.

XII.4. Experimental procedures

People working with highly stressed insulators may face two types of problems.

On the one hand, they may have to solve technical problems such as finding the highest voltage which can be applied to a given insulation to give a probability smaller than some value - say 10^{-3} - that breakdown will occur during a given time interval.

On the other hand, they may deal with apparently more academic questions such as the possible breakdown process which might take place in a given insulation subjected to a given field for a given time.

In fact, both types of problems are related, and an adequate understanding of the physical processes usually helps to solve the technical questions in a pragmatic way.

Consequently, measurements of dielectric strength should always be carried out with a scientific approach, and this requires:

(a) knowledge that the dielectric strength is not a unique, intrinsic material property.

(b) variation of the experimental parameters (sample thickness and preparation, electrodes etc.) and assessment of the relevance of each.

Naturally, the greatest care both in the choice of method and in the experimental work should be taken, and a few hints are given below.

1. Sample preparation

Although many standard techniques have been proposed by official Committees (I.E.C., I.S.O., C.I.G.R.E., A.S.T.M., etc.) the major principles for performing meaningful breakdown tests on solid samples are given below.

To avoid the use of very high voltages, it is advisable to test thin materials, preferably in the thickness range 0.1 to 1 mm. Errors in the thickness measurement may become significant with very thin samples, but a sample thickness exceeding 1 mm may require 200 kV or more for impulse breakdown to occur.

Such a voltage is not only dangerous and expensive to generate, but it gives rise to flashover in the surrounding medium, unless proper care is taken. This may involve shaping thick samples as shown in Fig. 61, forcing breakdown to take place in the thinner part. Multiple cups for repeated tests can be drilled in large samples.

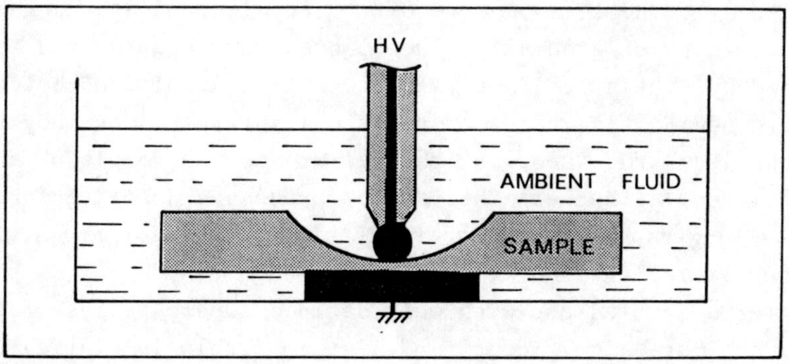

Fig. 61 - Schematic set-up for a breakdown test.

The best way to avoid flashover is to immerse the sample in a medium of high dielectric strength. Dehydrated transformer oil, silicon oil, etc. are very good insulators, but they often impregnate the material under test, and their wetting property might prove quite annoying. Electronegative gases, and in particular sulphur hexafluoride (SF_6) with a dielectric strength of nearly 100 kV/cm^{-1} and a high density which makes its handling similar to that of liquids, are quite convenient.

An important requirement which should always be fulfilled is perfect adhesion between the material under test and the electrodes, since air gaps favour the onset of partial discharges, which may develop into a premature breakdown spark. There are several ways to prepare satisfactory electrodes; one of the best and easiest, whenever the equipment is available, is thermal evaporation of a thin layer of gold, aluminium or silver.

For perfect adhesion of the layer, the sample should first be thoroughly cleaned with dehydrated organic solvents to remove all traces of dust, grease and moisture. Then, evaporation should be done rather quickly, in a good vacuum (10^{-6} Torr) to obtain a low resistance layer about 2000 Å thick.

The breakdown tests can be done either in a very good vacuum (10^{-6} Torr or better) or in dry air at atmospheric pressure, or in any of the dielectric fluids mentioned above. Of course, a poor vacuum is to be avoided, since gases break down at low voltages in the vicinity of the Paschen minimum.

2. Voltage application and statistical analysis

Industrial tests are made with a.c. at the network frequency (50 or 60 Hz). The r.m.s. value of the applied voltage is raised - either by steps or gradually - until breakdown occurs, and the data are processed by statistical analysis, as will be explained below. These tests are, evidently, of great practical interest, and they are described in the standards. However, the failure is the result of an extremely complex interplay between the various processes described above, together with the long-range degrading effect of partial discharges which may occur at each half cycle.

Tests carried out with well-defined square pulses of applied voltage are more significant, and give more help in clarifying the physical processes involved in high field conduction and breakdown. Furthermore, these tests are the only ones which yield the statistical time-lag, a most important quantity in the understanding of breakdown.

Let V_s be the lowest voltage which always produces breakdown shortly after it is applied. A first series of tests can be carried out with step functions of amplitude $V_1 > V_s$. In each test, breakdown occurs after a different delay (or time lag) τ, and the net result of a series of N similar tests with N identical samples can be presented in an histogram of the distribution of the time lags.

From this histogram, the nature of the distribution (for instance normal or Gaussian, Poissonian, etc.) can be deduced and sometimes interpreted qualitatively.

Moreover, quantitative information on the distribution of time-lags, namely:

(a) the average: $\bar{\tau}(V_1) = \frac{1}{N} \sum_{i=1}^{i=N} \tau_i(V_1)$

(b) the standard deviation:

$$\sigma_1 = \left[\frac{1}{N} \sum_{i}^{N} (\tau - \tau_i)^2 \right]^{1/2} = \sqrt{\overline{\tau^2} - \bar{\tau}^2}$$

and higher order "moments" of the distribution, can be obtained.

If another series of tests is made in a similar way, using a voltage $V_2 > V_1$, a distribution of time lags is obtained, with a new average:

$$\bar{\tau}(V_2) = \frac{1}{N} \sum_{i=1}^{i=N} \tau_i (V_2) < \bar{\tau}(V_1)$$

and a new standard deviation σ_2.

Provided that enough samples and time are available (at least 50 tests, in view of the dispersion, at 10 different voltages requiring at least 500 samples, representing about 100 hours of work), a curve such as Fig.62 may be drawn.

The shape of this curve, together with the nature and the voltage dependence of the time-lag distribution, and recordings of the initial step of the breakdown current constitute the most detailed experimental approach to breakdown.

This analysis has been and is being attempted mainly on fluids, which are easier to handle than solids, since fresh samples can be introduced in the cell by circulation, without moving the electrodes. However, in order to prevent the intense breakdown current from deteriorating the electrode surfaces, so that the experimental conditions of each test are not affected by the previous test, it is necessary to cut-off the applied voltage as soon as possible after breakdown initiates, and to divert into an external circuit any transient current which might flow after the cut-off.

In view of the enormous experimental problems raised by basic breakdown studies, it is not surprising that the physical insight of breakdown in condensed phases is as yet rather rudimentary. Except in very special cases (high purity semiconductors at low temperatures, and a few very thin films), the avalanche in the condensed phase has never been analyzed for comparison with a detailed model.

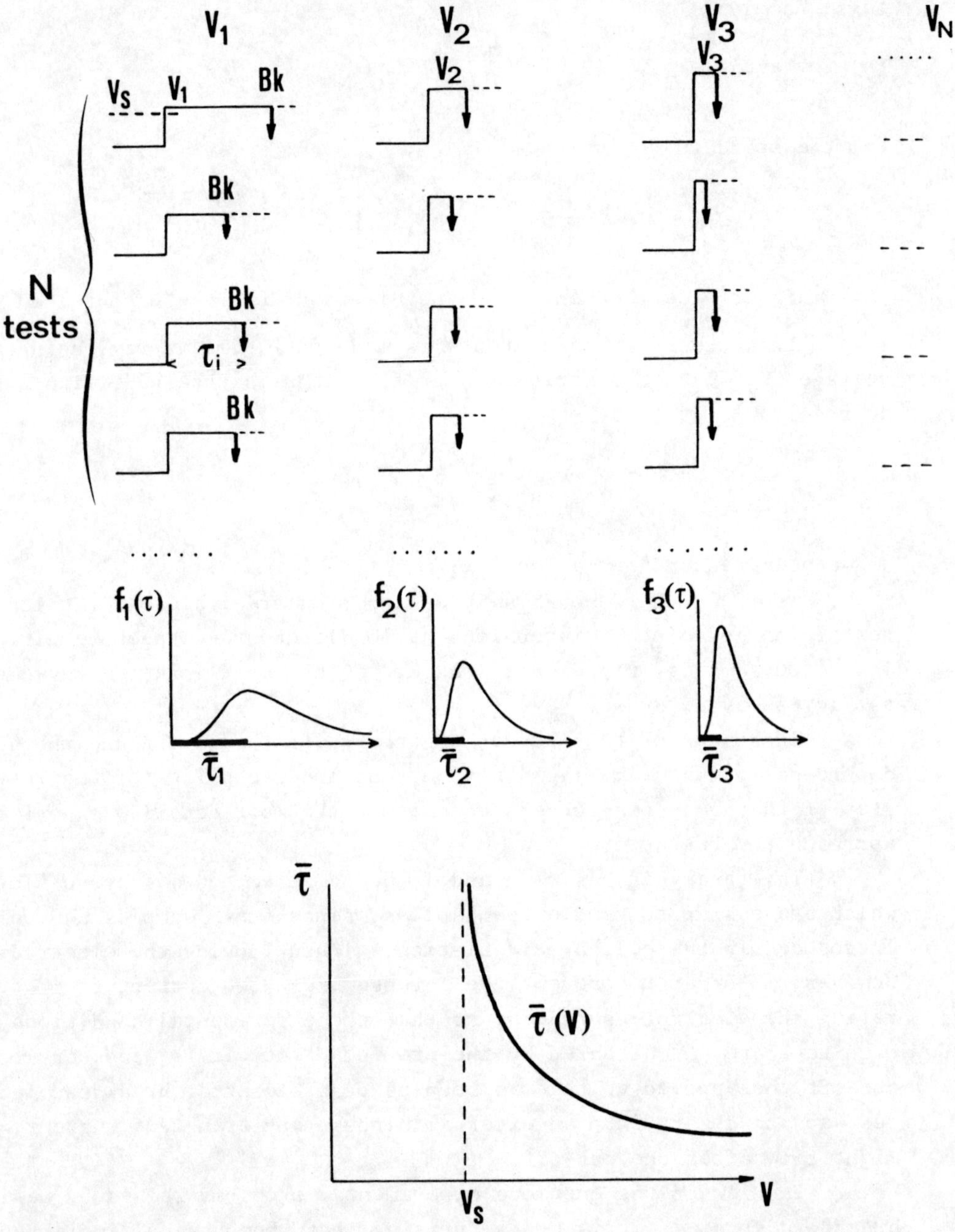

Fig.62 - Statistical determination of the breakdown time-lag as a function of the applied voltage.

In liquids, the bubbles and the fluid motion appearing with breakdown complicate an already intricate situation, and it is still a matter of dispute whether these bubbles are actually a cause or a consequence of breakdown.

In conclusion, the physics of the dielectric strength of condensed phases is an open field in which interdisciplinary talents are badly needed.

GENERAL BIBLIOGRAPHY

(*Stands for "review article")

BIBLIOGRAPHY ON PART 1

ELECTROSTATICS

W.R. SMYTHE — Static and Dynamic Electricity. McGraw-Hill, 1950.

L.D. LANDAU, E.M. LIFSCHITZ — Course of Theoretical Physics. Pergamon, 1960.

W.K.H. PANOVSKY, M. PHILIPS — Classical Electricity and Magnetism. Addison-Wesley, 1962.

N. FELICI — Electrostatique. Gauthier-Villars, 1962.

E. DURAND — Electrostatique, Vol. I, II and III. Masson, 1966.

PERMANENT AND INDUCED DIPOLE MOMENTS

P. DEBYE — Polar Molecules. Chemical Cat., 1929 ; Dover, 1947.

J.H. Van VLECK — Theory of Electric and Magnetic Polarizabilities. Oxford, 1932.

R.J. Le FEVRE — Dipole Moments. Methuen, 1948.

H. FRÖHLICH — Theory of Dielectrics. Clarendon, 1958.

C.J. BÖTTCHER — Theory of Electric Polarization. Elsevier, 1952, 2nd edn., Vol. 1, 1973, Vol. 2, 1978.

J.W. SMITH — Electric Dipole Moments. Butterworth, 1955.

A. Von HIPPEL — Dielectrics and Waves.
Wiley, 1954.

A. Von HIPPEL — Dielectric Materials and Applications.
Wiley, 1955.

J. BARRIOL — Les Moments Dipolaires.
Gauthier-Villars, 1957.

D.W. DAVIES — The Theory of the Electric and Magnetic Properties of Molecules.
Wiley, 1967.

L. EYRAUD — Dielectriques Solides Anisotropes.
Gauthier-Villars, 1967.

A.D. BUCKINGHAM* — Electric moments of molecules, in Physical Chemistry - Advanced Treatise, pp. 349-384.
Academic Press, 1970.

V.I. MINKIN et al. — Dipole Moments in Organic Chemistry.
Plenum, 1970.

A.A. MARYOTT — Tables of Dielectric Constants and Dipole Moments in the Gaseous Phase.
NBS Doc 537, 1953.

KERR EFFECT

Z. CROITORU* — Space-chargein dielectrics, in Progress in Dielectrics - Vol. 6, pp. 103-146 (Applications).
Heywood, 1965.

H.A. STUART — Molekülstruktür.
Springer, 1967.

C.G. Le FEVRE*
R.J. Le FEVRE — The Kerr effect, in Weissberger and Rossiter, Techniques of Chemistry, Vol. I, p. 399.
Wiley, 1972.

S. KIELICH — Electro-optical properties of dielectrics, in Dielectric and Related Molecular Processes, Vol. 1, pp. 278-278.
The Chem. Soc., London, 1972.

BIBLIOGRAPHY ON PART 2

DIELECTRIC DISPERSION IN SINGLE PHASES

C.P. SMYTH — Dielectric Behaviour and Structure.
McGraw-Hill, 1955.

W.F. FÜELLER BROWN	Dielectrics, in Handbuch der Physik, Vol. XVII, p. 1-154. Springer, 1956.
J.C. ANDERSON	Dielectrics. Chapman & Hall, 1964.
V. DANIEL	Dielectric Relaxation. Academic Press, 1967.
R.H. COLE*	Theories of dielectric polarization and relaxation, in Progress in Dielectrics, Vol. 3, pp. 47-100. Heywood, 1961.
R.J. MEAKINS*	Mechanisms of dielectric absorption in solids, in Progress in Dielectrics, Vol. 3, pp. 151-202. Heywood, 1961.
B.K.P. SCAIFE*	Dispersion and fluctuations in dielectrics, in Progress in Dielectrics, Vol. 5, pp. 1-186. Heywood, 1963.
J.J. O'DWYER* E. HARTING	Mechanisms of dielectric absorption in solids, in Progress in Dielectrics, Vol. 7, pp. 1-44. Heywood, 1967.
A. BELLEMANS	Statistical Theory of Electric Polarization Argonne National Laboratory - AEC. Publication 7381 (1967).
B.K. SCAIFE	Complex Permittivity. English Univ. Press, 1971.
G. WYLLIE*	Dielectric relaxation and molecular correlations, in Dielectric and Related Molecular Processes, Vol. 1, pp. 21-63. The Chem. Soc. London, 1972.
N. HILL, W.E. WAUGHAM, A.H. PRICE and M. DAVIES	Dielectric Properties and Molecular Behaviour. Van Nostrand, 1969.
A.M. NORTH*	Dielectric relaxation, in Chemical and Biological Applications of Relaxation Spectrometry. Reidel, 1975.
J. BERTIN J. LOEB	Experimental and Theoretical Aspects of Induced Polarization, Vols. 1 and 2. Lubrecht & Cramer, 1976.

A.K. JONSCHER* — The universal dielectric response, (R.A. + Contrib.). Nature, 267 (1977), 673.

HETEROGENEOUS DIELECTRICS

L.K.H. Van BEEK* — Dielectric Behaviour of Heterogeneous Systems, in Progress in Dielectrics, Vol. 7, pp. 69-114. Heywood, 1967.

BIBLIOGRAPHY ON PART 3

GENERAL

Proceedings of the 2nd International Conference on Conduction and Breakdown in Dielectric Liquids, Durham, 1963. J. Morant, Editor. Durham University, 1963.

I. ADAMCZEWSKI — Ionization, Conductivity and Breakdown in Dielectric Liquids. Taylor & Francis, 1969.

Dielectric Materials, Measurements and Applications. Proceedings of the Lancaster 1970 Conference I.E.E. Public., 67 (1970).

Phénomènes de conduction dans les liquides isolants. Proc. 3rd Int. Conf. on Dielectric Liquids, Grenoble, 1968. Editions du CNRS (1970).

R. COELHO* — Liquides dielectriques, in Techniques de l'Ingénieur, 6 (1971), D224.

P. HARROP — Dielectrics. Butterworth, 1972.

Proceedings of the 4th International Conference on Conduction and Breakdown in Dielectric Liquids, 1972. T.J. Gallagher, Editor. University College Press, Dublin, 1972.

A.A. ZAKY, R. HAWLEY — Conduction and Breakdown in Mineral Oil. Peregrinus, 1973.

R.W. SILLARS — Electrical Insulation Materials and their Application. Peregrinus, 1974.

Proceedings of the 1976 IEEE International Symposium on Electrical Insulation. Montreal, 1976.

T.J. GALLAGHER — Simple Dielectric Liquids Mobility, Conduction and Breakdown. Clarendon Press, 1975.

Proceedings of the 5th International Conference on Conduction and Breakdown in Dielectric Liquids. J.M. Goldschvartz, Editor. Delft University Press (Netherlands), 1975.

Proceedings of the International High Voltage Symposium, Zurich, 1975. Published in ETZ, A.97, 1976.

CONDUCTION

J.J. THOMSON — Conduction of Electricity through Gases. Cambridge University Press, 1928.

C. KITTEL — Introduction to Solid State Physics, 2nd edn. Wiley, 1956.

A. Van der ZIEL — Solid State Physical Electronics. Prentice-Hall, 1957.

A.I. GUBANOV — Quantum Electron Theory of Amorphous Conductors. Consultants Bureau, New York, 1965.

R.H. TREDGOLD — Space-charge Conduction in Solids. Elsevier, 1966.

W.R. HARPER — Contact Electrification. Clarendon Press, 1967.

F. GUTMANN, L.E. LYONS — Organic Semiconductors. Wiley, 1967.

J. van TURNHOUT — Thermally Stimulated Discharge of Polymer Electrets. Elsevier, 1975.

Electrets and Related Electrostatic Charge Storage Phenomena. Proceedings of the Chicago Symposium, 1967. M.L. Baxt and M.M. Perlman, Editors. The Electrochemical Society, 1968.

M.A. LAMPERT, P. MARK — Current Injection in Solids. Electrical Science Series. Academic Press, 1970.

N.F. MOTT E.A. DAVIS	Electronic Processes in Non-crystalline Materials. Clarendon Press, 1971.
J.G. SIMMONS	D.C. Conduction in Thin Films. Crane-Russak, 1971.
J.T. DEVREESE	Polarons in Ionic Crystals and Polar Semiconductors. North-Holland, 1972.
M.E. MILBERG	Fast Ion Transport in Solids. Nort-Holland, 1973.
P.D. TOWNSEND J.C. KELLY	Color Centers and Imperfections in Insulators and Semiconductors. Crane-Russak, 1973.
	Proceedings of the NATO Summer Course on Electronic Structure of Polymers and Molecular Crystals. J.M. André et al. Editors. Academic Press and NATO, 1975.
K.J. EULER*	Elektrete, Ein Uberblick über den Heutigen Stand. J. Electrostatics, 2 (1976), 1.
S. HUNKLINGER* H. SUSSNER K. DRANSFELD	New dynamic aspects of amorphous dielectric solids, in Advances in Solid-State Physics, Vol. 16, p. 267. Vieweg, 1976.
J. FUKRMANN*	Electrical properties of polymer solids in strong electric field, Colloid and Polymer Science, 254-2 (1976), 129.
R. GOFFAUX*	Sur les propriétés diélectriques des films de hauts polymères. Bulletin A.I.M. (Belgium), 2 (1976), 197.
J.R. MacDONALD*	Space charge polarization, in Electrode Processes in Solid State Ionics, M. Kleitz and J. Dupuy, Editors, p. 149. Reidel, 1976.
C.H.S. DUPUY A. CACHARD	Physics of Non-Metallic Thin Films. Plenum Press, 1976.
G. WEBER	Untersuchungen zum Leitungsmechanismus in Polyäthylen. Dissertation, Darmstadt, 1976.
R.E.B. BARKER*	Mobility and Conductivity of Ions in and into Polymeric Solids. Pure and Applied Chemistry, 46 (1976), 157.

BREAKDOWN

J.S. TOWNSEND	The Theory of Ionization of Gases by Collision. Constable, 1910.
A. NIKURADZE	Das Flüssige Dielektrikum. Springer, 1934.
J.M. MEEK J.D. CRAGGS	Electrical Breakdown in Gases. Clarendon Press, 1953.
W. FRANZ*	Dielektrischer Durchschlag, in Encyclopedia of Physics, Vol. 17 (Dielectrics), p. 155. Springer, 1956.
S. WHITEHEAD	Dielectric Breakdown in Solids. Clarendon, 1959.
T.J. LEWIS*	The Electric Strength and High-field Conductivity of Dielectric Liquids, in Progress in Dielectrics, Vol. 1, p. 97. Heywood, 1959.
J.H. MASON*	Dielectric Breakdown in Solid Insulation, in Progress in Dielectrics, Vol. 1, p. 1. Heywood, 1959.
G. WYLLIE*	Theory of polarization and absorption in dielectrics : an introductory survey, in Progress in Dielectrics, Vol. 2, p. 1. Heywood, 1960.
A.J. CURTIS*	Dielectric properties of polymeric systems, in Progress in Dielectrics, Vol. 2, p. 29. Heywood, 1960.
J.A. KOK	Electrical Breakdown of Insulating Liquids. Philips Technical Library, 1961.
R. STRATTON*	The theory of dielectric breakdown in solids, in Progress in Dielectrics, Vol. 3, p. 233. Heywood, 1961.
A.H. SHARBAUGH* P.K. WATSON	Conduction and breakdown in liquid dielectrics, in Progress in Dielectrics, Vol. 4, p. 199. Heywood, 1962.
R. COOPER*	The electric breakdown of alkali halide crystals, in Progress in Dielectrics, Vol. 5, p. 95. Heywood, 1963.
N. KLEIN*	Electrical breakdown in solids, in Advances in Electronics and Electron Physics, 26 (1969), 309-424.

A.A. ZAKY Dielectric Solids.
R. HAWLEY Solid State Physics Series.
Routledge & Kegan Paul, 1970.

J.J. O'DWYER The Theory of Dielectric Breakdown in Solids.
Clarendon, 1973.

CHARGE INJECTION IN VACUO AND GASES

Encyclopedia of Physics, Vol. 21, Springer 1956, contains majors articles on thermoionic emission (W. Nottingham), field emission (R.H. Good and E.W. Müller), secondary emission (R. Kollath), etc...

ANNUAL PUBLICATIONS AND SURVEYS

Annual Reports of the Conference on Electrical Insulation and Dielectric Phenomena.
National Research Council.
National Academy of Sciences.
Constitution Ave., Washington D.C.

Digest of Literature on Dielectrics (Yearly Digest).
National Academy of Sciences - National Research Council.

Progress in Dielectrics.
Volumes 1 (1959) to 7 (1967).
J.B. Birks, Editor.
Heywood, London.

ASTM Standards on Electrical Insulation.
Part 39 : Test Methods.
Part 40 : Specifications.
Philadelphia.

INDEX

Words frequently used in this book, such as "breakdown", "dipole", "polarization", etc... are not included in the Index. For reference to these words, the reader should consult the Table of Contents.